# BIG DATA ANALYTICS

Harnessing Data for New Business Models

# BIG DATA ANALYTICS

## Harnessing Data for New Business Models

*Edited by*
**Soraya Sedkaoui, PhD**
**Mounia Khelfaoui, PhD**
**Nadjat Kadi, PhD**

First edition published 2022

**Apple Academic Press Inc.**
1265 Goldenrod Circle, NE,
Palm Bay, FL 32905 USA
4164 Lakeshore Road, Burlington,
ON, L7L 1A4 Canada

**CRC Press**
6000 Broken Sound Parkway NW,
Suite 300, Boca Raton, FL 33487-2742 USA
2 Park Square, Milton Park,
Abingdon, Oxon, OX14 4RN UK

**Library and Archives Canada Cataloguing in Publication**

Title: Big data analytics : harnessing data for new business models / edited by Soraya Sedkaoui, PhD, Mounia Khelfaoui, PhD, Nadjat Kadi, PhD.

Names: Sedkaoui, Soraya, editor. | Khelfaoui, Mounia, editor. | Kadi, Nadjat, editor.

Description: First edition. | Includes bibliographical references and index.

Identifiers: Canadiana (print) 20210000074 | Canadiana (ebook) 2021009009X | ISBN 9781771889568 (hardcover) | ISBN 9781003129660 (ebook)

Subjects: LCSH: Management—Data processing. | LCSH: Industrial management—Decision making. | LCSH: Business planning—Statistical methods. | LCSH: Sustainable development. | LCSH: Big data.

Classification: LCC HD30.2 .B54 2021 | DDC 658.4/038—dc23

**Library of Congress Cataloging-in-Publication Data**

Names: Sedkaoui, Soraya, editor. | Khelfaoui, Mounia, editor. | Kadi, Nadjat, editor.

Title: Big data analytics : harnessing data for new business models / edited by Soraya Sedkaoui, PhD, Mounia Khelfaoui, PhD, Nadjat Kadi, PhD.

Description: First edition. | Palm Bay, FL : Apple Academic Press, 2021. | Includes bibliographical references and index.

Subjects: LCSH: Business--Data processing. | Business--Technological innovations. | Sustainable development--Data processing. | Big data.

Classification: LCC HF5548.2 .B4625 2021 (print) | LCC HF5548.2 (ebook) | DDC 658/.0557--dc23

LC record available at https://lccn.loc.gov/2020055946

LC ebook record available at https://lccn.loc.gov/2020055947

ISBN: 978-1-77188-956-8 (hbk)
ISBN: 978-1-77463-786-9 (pbk)
ISBN: 978-1-00312-966-0 (ebk)

# About the Editors

**Soraya Sedkaoui, PhD, HDR**
*Senior Lecturer, University Djilali Bounaama, Khemis-Miliana,*
*Algeria; Data Analyst and Strategic Business Consultant,*
*SRY Consulting Montpellier, France*

Soraya Sedkaoui, PhD, is a Senior Lecturer, Data Analyst, and Strategic Business Consultant with more than 10 years of teaching, training, research, and consulting experience in statistics, big data analytics, and machine learning algorithms. Leading the Big Data Analytic Consulting Practice at SRY Consulting in Montpellier, France, Dr. Soraya is focused on working with global clients across industries to determine how a data-driven approach can be embedded into strategic initiatives. This also includes helping businesses create actionable insights to drive business outcomes that lead to benefits valued in several fields. Dr. Soraya's works have contributed to delivering analytics services and solutions for competitive advantage through the use of algorithms, advanced analytical tools, and data science techniques. She worked as a researcher at TRIS Laboratory at the University of Montpellier, France (2011–2017). She contributed to the European project on "Internet Economics: Methods, Models, and Management (2017)" in collaboration with Pr. H-W Gottinger (STRATEC, Munich, Germany). She also contributed to creating many algorithms for business applications, such as the algorithm of Snail 2016, in France and more. Her science-oriented research experience and interests are in the areas of big data, computer science, and the development of algorithms and models for business applications and problems. Dr. Sedkaoui's prior books and research have been published in several refereed editions and journals. Dr. Soraya also holds a PhD in economic analysis and an HDR in economic and applied statistics.

**Mounia Khelfaoui, PhD, HDR**
*Teacher-Researcher and Lecturer, University Djilali Bounaama*
*Khemis-Miliana, Algeria*

Mounia Khelfaoui, PhD, is a teacher-researcher and Lecturer at the University Djilali Bounaama Khemis-Miliana in Algeria. With experience in research, she is a member of the research laboratory "Industry, Organizational Development of Enterprises and Innovation" of the University of Khemis-Miliana since 2008. Her research focuses on sustainable development, especially corporate social responsibility (CSR), the sharing economy, and the circular economy. She has published in various journals and conferences dealing with the topic of CSR and sustainable development. Dr. Khelfaoui's research proposes to demonstrate the role of the adoption of the CSR in organizations in light of the principles of sustainable development. She graduated from the University of Algiers 3 with a PhD in economics and an HDR in environmental economics.

**Nadjat Kadi, PhD, HDR**
*Senior Lecturer, University of Djilali Bounaama Khemis-Miliana,*
*Algeria; Manager, The Digital Economy Laboratory*

Nadjat Kadi, PhD, is a Senior Lecturer at the University of Djilali Bounaama Khemis-Miliana, Algeria. She is the Manager of The Digital Economy Laboratory. Her research relates to economic and statistical analysis and the field of demography. She graduated from the University of Oran, Algeria, with a PhD in demography and an HDR in economic and demographic analysis.

# Contents

# Contributors

**Abdellah Aggoun**
Assistant Professor, Faculty of Economics, Business, and Management Sciences, University of Djilali Bounaama, Khemis Miliana, Algeria, Rue Thniet El Had, Khemis Miliana, Ain Defla, Algeria, E-mail: agg88abd@gmail.com

**Malika Bakdi**
Senior Researcher, National High School of Statistics and Applied Economics (ENSSEA), Koléa, Algeria, E-mail: bakdi_malika@yahoo.fr

**Houssame Eddine Balouli**
National High School of Statistics and Applied Economics (ENSSEA), Koléa, Algeria, E-mail: balouli.houssame.eddine@gmail.com

**Hamza Belghalem**
Temporary Assistant Professor, Faculty of Economics, Business, and Management Sciences, University of Djilali Bounaama, Khemis Miliana, Algeria, E-mail: hamzabelghalem44@gmail.com

**Khedidja Belhadji**
PhD student, Specialization in Production Management, Hassiba Benbouali University, Chlef, Algeria, E-mail: nafoula80@gmail.com

**Khalida Mohammed Belkebir**
HDR, Senior Lecturer, and Researcher, Faculty of Business Economics and Management, Djillali Bounaama University, Theniet El Had Street, Khemis Miliana, W. Ain-Defla, Algeria, E-mail: k.mohammed-belkebir@univ-dbkm.dz

**Rafika Benaichouba**
PhD in Economic Sciences, Senior Lecturer, University of Djillali Bounaama, Khemis Maliana, Algeria, E-mail: benaichoubarafika@yahoo.fr

**Amal Bensautra**
PhD Student, Faculty of Economics, Business, and Management Sciences, University of Djilali Bounaama, Khemis Miliana, Algeria, E-mail: bensautra.amal@hotmail.com

**Rabia Ahmed Benyahia**
University of Djilali Bounaama, Khemis Miliana, Algeria, E-mail: rabiebenyahia33@yahoo.com

**Bakhta Bettahar**
University of Abdelhamid Ibn Badis, Mostaganem, Algeria, E-mail: bakhta_48@hotmail.fr

**Mustapha Bouakel**
Associate Professor, Faculty of Economics, Commerce, and Management Sciences, University Center Ahmed Zabana, Relizane, Algeria, E-mail: mustapha.bouakel@univ-sba.dz

**Wassila Chadli**
National High School of Statistics and Applied Economics (ENSSEA), Koléa, Algeria E-mail: chadli.wassila@outlook.com

**Lazhar Chine**
Associate Professor, Boumerdes University, Algeria, E-mail: l.chine@univ-boumerdes.dz

**Abdelkader Dahman**
Assistant Professor, Faculty of Economics, Business, and Management Sciences,
University of Djilali Bounaama, Khemis Miliana, Algeria, E-mail: abd19dah@gmail.com

**Dehbia El Djouzi**
Senior Lecturer and Researcher, Faculty of Business Economics and Management,
Djillali Bounaama University, Theniet El Had Street, Khemis Miliana, Algeria,
E-mail: raison81@yahoo.fr

**Redouane Ensaad**
Department of Commercial Sciences, University of Hassiba Ben Bouali, Chlef, Algeria,
Pb. 02000, Algeria

**Amel Fassouli**
PhD Student, Faculty of Economics, Business, and Management Sciences,
University of Djilali Bounaama, Khemis Miliana, Algeria, E-mail: Amelsabrine2018@gmail.com

**Rabah Ghazi**
Laboratory of Globalization, Politics, and Economics, University of Algiers 3, Dely Brahim, Algeria,
E-mail: ghazi.rabah@univ-alger3.dz

**Fella Ghida**
Senior Lecturer, and Researcher, Faculty of Economics, Business, and Management Sciences,
University of Djilali Bounaama, Khemis Miliana, Algeria, E-mail: fghida@yahoo.fr

**Saliha Hafifi**
Senior Lecturer, Faculty of Economics, Business, and Management Sciences,
University of Djilali Bounaama, Khemis Miliana, Algeria, E-mail: hafifis18@yahoo.fr

**Djazia Hassini**
Department of Economic Sciences, University of Hassiba Ben Bouali, Chlef, Algeria

**Fatima Zohra Hennane**
PhD Student, University Ali Lounici-Blida 2, Route d'El Afroun, Blida, Algeria,
E-mail: Hennane_fz@yahoo.fr

**Mohamed Ilifi**
Senior Lecturer, Faculty of Economics, Business, and Management Sciences,
University of Djilali Bounaama, Khemis Miliana, Algeria, E-mail: m.ilifi@univ-dbkm.dz

**Abdellah Kelleche**
Senior Lecturer, Hassiba Benbouali University, Chlef, Algeria, Pb. 02000, Algeria,
E-mail: kabd.dz@gmail.com

**Fatma Zohra Khebazi**
Lecturer, Faculty of Economics, Business, and Management Sciences, Khemis Miliana University,
Rue Thiniet El Had, Khemis Miliana, Ain Defla, Algeria, E-mail: fkhebazi@gmail.com

**Djamila Cylia Kheyar**
PhD Student, Faculty of Economics, Business, and Management Sciences,
University of Djilali Bounaama, Khemis Miliana, Algeria, E-mail: kheyar.djamilacylia@gmail.com

**Zahia Kouache**
Senior Lecturer, Faculty of Economics, Business, and Management Sciences,
University of Djilali Bounaama, Khemis Miliana, Algeria, E-mail: z.kouache@univ-dbkm.dz

**Fatima Lalmi**
PhD in Economic Sciences, Senior Lecturer, University of Abdelhamid Ibn Badis, Mostaganem,
Algeria, E-mail: lalmi.fatima@yahoo.fr

**Fethia Benhadj Djilali Magraoua**
Senior Lecturer, Faculty of Economics, Business, and Management Sciences,
University of Djilali Bounaama, Khemis Miliana, Algeria, E-mail: magr_fati@yahoo.fr

**Kamel Maiouf**
Department of Sciences Economy, Commercial, and Management Sciences,
Hassiba Ben Bouali University of Chlef, Pb. 02000, Algeria, E-mail: m.kamel@univ-chlef.dz

**Fatima Mana**
Senior Lecturer, Department of Management Sciences, University of Hassiba Ben Bouali,
Chlef, Algeria, E-mail: f.mana@univ-chlef.dz

**Zineb Matene**
Assistant Professor and Researcher, Faculty of Business Economics and Management,
Djillali Bounaama University, Theniet El Had Street, Khemis Miliana, W. Ain-Defla, Algeria,
E-mail: z.matene@univ-dbkm.dz

**Nachida Mazouz**
Lecturer, Faculty of Economics, Business, and Management Sciences, University Ali Lounici-Blida 2,
Route d'El Afroun, Blida, Algeria

**Nadia Messaoudi**
Assistant Professor, Faculty of Economics, Business, and Management Sciences,
University of Djilali Bounaama, Khemis Miliana, Algeria, E-mail: n.messaoudi@univ-dbkm.dz

**Achour Mezrig**
Department of Sciences Economy, Commercial, and Management Sciences,
Hassiba Ben Bouali University of Chlef, Pb. 02000, Algeria, E-mail: m.kamel@univ-chlef.dz

**Boulanouar Mokhtari**
Senior Lecturer, Faculty of Economics, Business, and Management Sciences,
University of Djilali Bounaama, Khemis Miliana, Algeria, E-mail: b.mokhtari@univ-dbkm.dz

**Nadia Hamdi Pacha**
Lecturer and Researcher, Faculty of Economics, Business, and Management Sciences,
University Ali Lounici-Blida 2, Route d'El Afroun, Blida, Algeria,
E-mail: hamdipacha.n@hotmail.com

**Fatima Rachedi**
University of Larbi Ben Mhidi, Oum El Bouagui, Algeria, E-mail: rachedi.fatima@yahoo.fr

**Khadra Rachedi**
University of Oran 2, Algeria, E-mail: Kha-dra@hotmail.fr

**Djamila Sadek**
Faculty of Economics, Business, and Management Sciences, University Center of Tissemsilt, Algeria,
E-mail: sadek.djamila.ecom@gmail.com

**Ramdhan Sahnoun**
PhD Student, Faculty of Economics, Business, and Management Sciences,
University of Djilali Bounaama, Khemis Miliana, Algeria, E-mail: laz152.rs@gmail.com

**Fatima Zohra Soukeur**
Laboratory of Globalization, Politics, and Economics, University of Algiers 3, Dely Brahim, Algeria,
E-mail: zola_marketing@yahoo.fr

**Nadia Soudani**
Faculty of Economics, Business, and Management Sciences, University Center of Tissemsilt, Algeria,
E-mail: soudani_mag@hotmail.com

**Yahia Benyahia**
PhD Student, University Ali Lounici-Blida 2, Route d'El Afroun, Blida, Algeria,
E-mail: ey.benyahia@univblida2.dz

**Noureddine Zahoufi**
Assistant Professor, Faculty of Economics, Business, and Management Sciences,
University of Djilali Bounaama, Khemis Miliana, Algeria, E-mail: zahoufi.norddine@gmail.com

**Amina Zerbout**
PhD Student, Faculty of Economics, Commerce, and Management Sciences, University Ali Lounici,
Blida 2, Algeria, E-mail: ea.zerbout@univ-blida2.dz

# Abbreviations

| | |
|---|---|
| ACM | Association for Computing Machinery |
| ATIH | Technical Agency for Hospital Information |
| AWS | Amazon web services |
| BD | big data |
| BI | business intelligence |
| BSP | bulk synchronous parallel |
| CDC | Centers for Disease Control and Prevention |
| DBMS | definition of the table in the system |
| DM | data mining |
| EFA | Education For All |
| ETL | extraction, transformation, and loading |
| FG-SSC | focus group on sustainable smart cities |
| GPS | global positioning system |
| HDFS | Hadoop distributed file system |
| HIS | hospital information system |
| ICTs | information and communication technologies |
| IDC | International Data Corporation |
| IoT | internet of things |
| ISO | International Standards Organization |
| ITC | information, technology, and communications |
| JSON | JavaScript object notation |
| ML | learning models |
| PCI | payment card industry |
| PMSI | Programme de Médicalisation des Systèmesd'Information |
| RDD | resilient distributed datasets |
| RFID | radio-frequency identification |
| SMA | social media analytics |
| SMEs | small and medium enterprises |
| URL | uniform resource language |
| WCED | World Commission on Environment and Development |
| WHO | World Health Organization |

# Preface

*"Where there's data smoke, there's business fire."*
**—Thomas C. Redman**
*Data-Driven: Profiting from Your Most Important Business Asset*

In recent years, significant investments have been made in companies' infrastructure to increase their data collection capacity. Practically, all aspects of a business are now open to data collection: operations, manufacturing, supply chain management, customer behavior, the performance of marketing campaigns, flow management procedures, etc.

Simultaneously, data about events outside the company, such as market trends, company news, and competitors' activities, is now widely available. This data availability has sparked a growing interest in methods of extracting useful information and knowledge from data: the field of "big data analytics."

Big data and data analytics are being adopted more frequently, especially in companies looking for new methods to develop smarter capabilities and tackle challenges in the dynamic processes. The possible uses of big data analytics are numerous and cross-sector. With the vast amounts of data available today, companies in every sector are now focusing on harnessing data to create a new way of doing business.

The current discussion about this field, which is often referred to as revolutionary, can be described using W. Edwards Deming's description:

> Data are not taken for museum purposes; they are taken as a basis for doing something. If nothing is to be done with the data, then there is no use in collecting any. The ultimate purpose of taking data is to provide a basis for action or a recommendation for action. The step intermediate between the collection of data and the action is prediction.

In addition, due to big data analytics' cross-business application scenarios, several specific business concepts are also affected. The analysis, therefore, focuses on both technical and organizational aspects of big data tools and technologies.

Therefore, the challenges of the current business playground require a radical change in the manner of exploring the potential associated with data for creating value, which presents a pillar of business sustainability nowadays.

In this context, the 4th National Conference on *"Big Data Analytics: Harnessing Data for New Business Models"* (BDA2019) aimed to provide a forum for researchers alike to exchange the latest fundamental advances in the big data field and its best practices, and as well as emerging research topics that would define the future of big data applications in the business context.

BDA2019 emerged as an outcome of several research results from Algerian academics to provide relevant lessons learned from specific data uses that generate value in the business context. During the 1st and 2nd of October 2019, at the Faculty of Economics at the University of Khemis Miliana, Algeria, we have celebrated and shared the knowledge on this exciting field. In these two special days, the BDA2019 has provided researchers, academics, and experts an opportunity to exchange and share their research experiences and results and deepen the debate on data-driven value creation.

This conference aimed to work out possible potentials based on a basic introduction to big data analytics, before the main sections dealt in detail with the challenges relating to this innovative technology, its diverse applications in the business context, how this technology enhances the decision-making process, and how it contributes to achieving the sustainable development goals.

But the raised exciting question of BDA2019 was the application of the advanced tools and technologies of this emerging field and its evidential value within businesses around the world. The question is whether businesses accross the world will adapt to this paradigm or whether the big data can be integrated into the architecture of global business.

This book gathers selected works related to big data applications in several areas, focusing on the diverse points discussed during these two business days. Throughout this book's four parts, we will detail various subjects and techniques relating to big data analytics and its applications.

We hope this book can encourage more engaging research at national and international levels on the big data applications in the business context. We wish you an exciting and stimulating reading and formulate the necessary bases to resolve big data dilemmas in business practice!

*—Editors*

# Acknowledgments

We are pleased to thank the authors whose submissions and participation made this conference possible. We also want to express our thanks to the Program Committee members for their dedication in organizing the conference. Also, we would like to thank Apple Academic Press (AAP) team for their help during the editing process of this book, especially Sandra Jones Sickels, Ashish Kumar, Sheetal, and Rakesh. Finally, the reviewers for their hard work reviewing process, which was essential for the success of BDA2019 and the publication of this book.

*—Soraya, Mounia, and Nadjat*
*Editors*

# PART I

# Big Data: Opportunities and Challenges

# CHAPTER 1

# Big Data: An Overview

MALIKA BAKDI[1] and WASSILA CHADLI[2]

[1]*Senior Researcher, National High School of Statistics and Applied Economics (ENSSEA), Koléa, Algeria, E-mail: bakdi_malika@yahoo.fr*

[2]*National High School of Statistics and Applied Economics (ENSSEA), Koléa, Algeria, E-mail: chadli.wassila@outlook.com*

## ABSTRACT

This chapter focuses on a new trend to process and analyze large data, i.e., big data. It has become an imperative approach, particularly with the massive outbreak of data on the Internet (videos, photos, messages, social networks, e-commerce transactions, etc.) and the large diffusion use of connected objects (smartphones and tablets). In this research, we attempt to represent the big data phenomenon's designs, architectures, and applications.

## 1.1 INTRODUCTION

Data and algorithms shape a new world that consists of a form of culmination for computing and, more precisely, a new way of controlling information. With more than 95% of the world's data set having been created in recent years, it is important to know that it is not the one who has the best algorithm wins, but the one who has more data; and it is not just any type of data, but only the reliable data that are counted. As a result, a large amount of data will be accumulated as we have algorithms that work very efficiently based on the data we process.

Thus, the major problem with this large amount of data is that it becomes very difficult to work with, especially with the traditional database processing tools [4]. Today, companies are facing an exponential increase in data volume. To give us a more precise idea, we can attain several petabytes $(10)^{15}$, see even zettabytes $(10)^{21}$.

As expected, the amount of data created and managed has grown exponentially over the past few years. Hence, we can imagine how huge the amount of data that will be created in the future years, as data can be acquired from logs, social media, e-commerce transactions (the data are of a diverse nature), etc. Undoubtedly, many companies want to take advantage of this data – whether data collected by themselves or public data such as the web or open data. As a result, traditional technologies are not designed to process with a massive data explosion, and therefore thanks to big data, where the exponential growth of data can be processed.

In this work, we present theoretical research about big data. It should be mentioned that 2012 was the year of the big data buzz when the notion was popularized; this means that companies are dealing with an amount volume of data to be processed, which presents a technical and economic challenge.

The objective of the present work is to answer the following questions: what is big data? Why are we interested in big data? In addition, what is the revolutionary technology adopted by big data?

## 1.2   BIG DATA: CONCEPT AND DEFINITION

Certainly, in the explanation of big data, a lot has been said about the volume, which is one of the very important aspects of the clarification of the big data concept. Thus, a classic definition has been proposed by Gartner, which implies three dimensions (as shown in Figure 1.1).

**FIGURE 1.1**   The three V's of big data.
*Source:* Authors' creation.

The first one is about *volume*: it is the massive explosion of data that requires their processing and analysis. The second dimension is *variety*, which corresponds to the difficulty of processing and analyzing data, but more precisely, crossing the new data sources in an effective way that is more diverse and from multiple nature. Thus, the variety distinguishes big data from traditional data analysis. Indeed, big data analyzes data sets from different sources [8]. The third dimension is the *velocity*, which corresponds to the speed with which they are generated, processed, and stored.

It is clear that individuals and companies are great data generators in a very short time, but there is a shifted time between their processing and their generation. The coming of big data technology makes the job easier, thus giving us the advantage of processing data while it is being generated.

Subsequently, the explanation of big data does not focus exclusively on these three dimensions, as IBM has added two other dimensions to properly target the explanation, which are *veracity* and *value*. Veracity is the ability to have reliable data; for example, the generation of data by spambot is an example worthy of confidence. Another example is that of Mexico, where the presidential elections were made by a fake Twitter account.

The fifth is the *value*, having an equivalent meaning that the big data approach only makes sense to achieve strategic objectives related to individuals and the company, for the purpose of creating an added value, regardless of the field of activity. Thus, the success of a big data project is largely correlated by the creation of added value and new knowledge. The explanation of big data extends to the other 5V to note: validity, vulnerability, volatility, visualization, and variability.

## 1.3 BIG DATA IN DIGITS

One of the fundamental reasons for the existence of the big data phenomenon is the current extent to which information can be generated and made available [5]. The speed growth of data, especially those approved by intelligent objects, will reach more than 50 billion in the world in 2020. According to predictions, 40,000 billion data will be generated [14].

It is estimated that 90% of the data collected since the beginning of humanity have been generated only over the last two years, in which 70% of the data are created by individuals, although it is the companies that store and manage 80% of it.

Following this exponential trend in data, the countries became aware of the importance of big data, and thus in 2012, the U.S. announced a donation of 200 million dollars for research related to the theme of big data. In parallel, the big data strategy generates profits of $8.9 billion, which is the revenue generated by the big data market in 2014. Certainly, Amazon would generate 30% of its revenues through cross-selling [12].

## 1.3.1 BIG DATA ORIGIN

According to Fermigier [6], big data comes in particular from:

- **The Web:** Access logs, social networks, e-commerce, indexing, storage of documents, photos, videos, linked data, etc. (e.g., Google processed 24 petabytes of data per day with MapReduce in 2009).
- **The Internet and Connected Objects:** RFID, sensor networks, telephone call logs.
- **Science:** Genomics, astronomy, subatomic physics (e.g., the German Climate Research Centre manages a database of 60 petabytes).
- **Business:** e.g., Transaction history in a chain of hypermarkets.
- **Personal Data:** e.g., Medical records.
- **Public Data:** Open data.

## 1.3.2 BIG DATA PIONEERS

The massive growth of new big data technologies has become essential for many companies wishing to better know their suppliers and customers. The booming big data market includes several actors offering specific services [7].

Major web stakeholders, including Yahoo and Google search engines, as well as social media such as Facebook, also offer big data solutions. From 2004, Google proposed MapReduce, an algorithm capable of processing and storing a large amount of data. In 2014, Google announced its replacement by Google Cloud Dataflow, a SaaS solution.

Yahoo, for its part, is one of the main contributors to the Hadoop project by hiring Doug Cutting, its creator. The search engine has also created Horton works, a company dedicated entirely to the development of Hadoop.

Amazon, the American online retail giant, is also one of the pioneers of big data. Since 2009, it has provided companies with tools such as Amazon

Web Services (AWS) and Elastic MapReduce, better known as EMR. The latter is accessible to everyone since its use does not require any skill in installing and adjusting Hadoop clusters [8].

Everyday users and individuals produce a massive amount of data. This data presents many opportunities for companies. Big data is the largest volume of data that translates into the creation of new technology that facilitates the growth and development of big data, which can be broadly categorized into two main families.

On the one hand, storage technologies are driven particularly by the deployment of cloud computing. On the other hand, the arrival of adjusted processing technology, especially the development of new databases adapted to unstructured data (Hadoop) and the implementation of high-performance computing modes (MapReduce). Figure 1.2 summarizes the main technologies that support the deployment of big data.

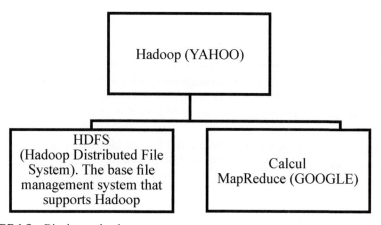

**FIGURE 1.2**   Big data technology.

## 1.4   BIG DATA ANALYTICS TYPES

The following four types of big data analytics were distinguished [9] (Figure 1.3):

- **Descriptive Analytics:** It consists of asking the question: "What is happening?" It is a preliminary stage of data processing that creates a set of historical data. Data mining (DM) methods organize data and help uncover patterns that offer insights.

- **Diagnostic Analytics:** It consists of asking the question: "Why did it happen?" Diagnostic analytics look for the root cause of a problem. It is used to determine why something has happened. This type attempts to find and understand the causes of events and behaviors.
- **Predictive Analytics:** It consists of asking the question: "What is likely to happen?" It uses past data in order to predict the future. It is all about forecasting. Predictive analytics uses many techniques such as DM and artificial intelligence to analyze the current data and make scenarios of what might happen.
- **Prescriptive Analytics:** It consists of asking the question: "What should be done?" It is dedicated to finding the right action to be taken. With descriptive analytics providing historical data and predictive analytics, helping forecast what might happen, prescriptive analytics use these parameters to find the best solution.

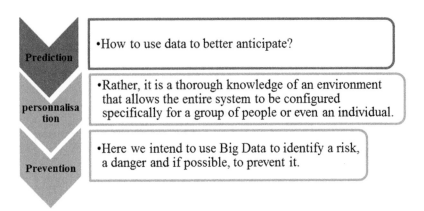

**FIGURE 1.3**   The 3Ps that describe big data purpose.

If the people now had not been living in an era where they produce a lot of data, the important question to ask would have been, "Will they adopt a big data approach?" The answer can be summarized in three main reasons:

First, the exponential increase in the number of connected users, connected smartphones, connected tablets, connected glasses, and as a result, connected objects. In addition, the individuals have become more reliant on terms of quality and costs. Finally, if so much data is being produced, data can be stored in different storages, especially with the digitization of society.

Certainly, big data plays a very important role for governmental organizations, private and multinational companies, whatever their field of activity, it applies to all types of companies, large or small, but with a necessary condition: it has to generate large volumes of data.

At first, big data was used by a specific sample of companies such as banks for credit card transactions and financial market-related uses, by telephone companies for telephone call records, and by e-commerce sites (e.g., Amazon and eBay) to improve online services. Although big data started in specific industries, it is now available to everyone, even small SMEs [10].

The value chain, the concept introduced by Porter [16], refers to a set of activities carried out to create added value at each stage of product design or to provide a service to its customers. Similarly, the data value chain refers to the framework that deals with a set of activities aimed at creating value from available data. It can be divided into four essential phases: data integration, data storage, data manipulation, data security, data analysis, and decision-making [11].

### 1.4.1   BIG DATA CLASSIFICATIONS

Big data can be classified into the following three categories [12]:

- **Structured Data:** It refers to any kind of data which is stored in relational databases and spreadsheets that reside in a fixed field within a record or file.
- **Unstructured Data:** The phrase unstructured data usually refers to information that doesn't reside in a traditional row-column database. As you might expect, it's the opposite of structured data-the data stored in fields in a database.
- **Semi-Structured Data:** It is data that hasn't been organized into a specialized repository, such as database, but even so, has associated information such as metadata, which makes it more amenable for processing than raw data.

The success of a big data project is largely linked by its architecture and its correct infrastructure, so the big data architecture is based on four components, as mention in Figure 1.4.

To summarize, an integration which consists of loading the volume of data onto storage media and then storing them in order to manipulate them, including the processing objective and better extract a reliable and correct result [13].

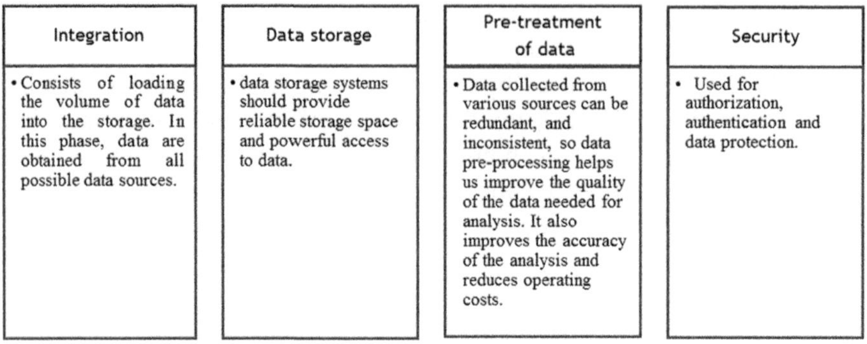

| Integration | Data storage | Pre-treatment of data | Security |
|---|---|---|---|
| • Consists of loading the volume of data into the storage. In this phase, data are obtained from all possible data sources. | • data storage systems should provide reliable storage space and powerful access to data. | • Data collected from various sources can be redundant, and inconsistent, so data pre-processing helps us improve the quality of the data needed for analysis. It also improves the accuracy of the analysis and reduces operating costs. | • Used for authorization, authentication and data protection. |

**FIGURE 1.4**    Big data architecture.

## 1.5 BIG DATA STRATEGY AND CHALLENGES

### 1.5.1 STRATEGY

The strategy is composed of five phases that involve different activities [7]:

- The hardware analysis is required for installing the software and the data to be analyzed, with the recommendation of a data server with a large storage capacity.
- The selection of the company's processes that will be analyzed can be customer sales processes, production data, equipment failures, among others; this process selection collects the necessary information and data that will be the raw material for the subsequent activities.
- The installation and configuration of the Hadoop platform are distributed data processing, as well as the software to support the Hadoop system.
- The extraction, transformation, and loading (ETL) activities with analysis services.
- The big data analytics, tools for analyzing reports (reporting), queries, and visualization (dashboards) will lead to data analytics.

### 1.5.2 CHALLENGES

The application of mass data has considerable benefits for individuals and society, but it also raises serious concerns about its potential impact on the dignity, rights, and freedom of the persons concerned, including their right to privacy.

These risks and challenges have already been the subject of multiple analyses by data protection specialists around the world. We can identify the two following concerns [14]:

1. **Lack of Transparency:** As the complexity of data processing increases, organizations often claim secrecy about how data are processed for reasons of commercial confidentiality. As in 2014, the White House Report noted, "some of the most important challenges revealed by this review are how massive data analysis can create a decision-making environment so opaque that individual autonomy disappears into an impenetrable set of algorithms" [15]. Unless natural persons receive appropriate information and have adequate control, individuals cannot exercise effective control over their data and give informed consent when required. This is particularly true for the precise future purposes of any secondary use of the data that may not be known at the time of data collection. In this case, the controllers may not be able or willing to explain to the data subjects precisely what will happen to their data and obtain their consent, if necessary.

2. **The Information Imbalance:** Between the organizations holding the data and the data subjects whose data they process is likely to increase with the development of applications based on massive data [14].

It should be mentioned that big data also have several weaknesses, such as:

- Detecting data judged abusive or earlier all data that will not follow a dominant statistical model, and we systematically remove any data contrary to the dominant statistical law.
- The absence of the quality of being reliable about results, the big data has the farcical tendency indeed to inspire and process a maximum of data, but without making a quantitative sorting. Here we can mention the famous example of the giant of the web Google, in 2011 to adopt a project to call Google flu trend to make a study on the evolution and the appearance of the influenza epidemic using its algorithm developed; they can collect data input on search engines as keywords,

for example, cough, flu, fever. However, the result was ambiguous and overestimated.

- The difficulty of processing what has not been detected and anticipated, and this makes us a tool that little performs at the novelty and breakdown.

## 1.6 CONCLUSION

The rise of big data is changing our world. In this chapter, we summarized the big data definition, characteristics (volume, variety, velocity, etc.), opportunities, and challenges. We noticed that the advent of big data technologies had been treated as a comparative advantage for professionals, in parity with companies generating large volumes of data that had difficulties in processing them. These advantages are present in the business's activity or business sector.

Therefore, we can say that big data makes it possible to get done analyses in real-time, predict, in the same way, to find solutions.

## KEYWORDS

- **big data**
- **big data architecture**
- **big data pioneers**
- **big data strategy**
- **digitization**
- **Hadoop system**

## REFERENCES

1. Amal, A., (2018). *Big Data*. National School of Engineers of Sfax (ENIS). France documents. https://fdocuments.fr/document/cours-big-data-chap1.html (accessed on 24 November 2020).
2. Jean-Pierre, R., & Floriane, D. K., (2016). *Big Data in Brussels Today and Tomorrow?* (p. 11). Les Cahiers D'Evoliris.
3. Andrea, D. M., Marco, G., & Michele, G., (2015). What is big data? A consensual definition and a review of key research topics. *AIP Conference Proceedings*, 98.

4. Thierry, B., (2017). *Journey into Big Data* (p. 64). Building together a sustainable digital trust, Voices of research. Clefs.
5. Sophie, D., (2014 & 2015). *Big Data Guide*. The reference directory.
6. Stefane, F., (2012). *Big Data and Open Source: An Inevitable Convergence.* Version 1.0.
7. Alicia, V., Griselda, C., et al., (2019). Big data strategy. *(IJACSA) International Journal of Advanced Computer Science and Applications, 10*(4).
8. *MBA ESG, Who are the Players in the Big Data Market?* https://www.mba-esg.com/actus/acteurs-big-data (accessed on 22 October 2020).
9. Youssra, R., (2018). Big data and big data analytics: Concepts, types, and technologies. *International Journal of Research and Engineering, 5*(9), 524–528.
10. Ali, K., (2011). *Qu'est-Ce Que Le Big Data (Big Data)? Big Data, Big Business.* https://kinaze.org/qu-est-ce-que-le-big-data-bigdata-definition/ (accessed on 22 October 2020).
11. Abhay, K. B., & Dhanya, J., (2017). *Big Data: Challenges, Opportunities, and Realities.*
12. Srinuvasu, M., Koushik, A., & Santhosh, E. B., (2018). Big data: Challenges and solutions. *International Journal of Computer Sciences and Engineering, 5*(10).
13. *Jean-Privat Desire BECHE, Massive Data Generalities: Big Data.* https://www.supinfo.com/fr/Default.aspx (accessed on 22 October 2020).
14. European Data Protection Supervisor (2015). *Meeting Big Data Challenges.* Avis n, 7.
15. Interim Progress Report, (2014). *Big Data: Seizing Opportunities, Preserving Values* (p. 10). Bureau exécutif du Président, mai.
16. Porter, M. E., (1980). *Competitive Strategy*. Free Press.

# CHAPTER 2

# Big Data between Pros and Cons

DJAMILA CYLIA KHEYAR

*PhD Student, Faculty of Economics, Business, and Management Sciences, University of Djilali Bounaama, Khemis Miliana, Algeria, E-mail: kheyar.djamilacylia@gmail.com*

## ABSTRACT

Big data is nowadays considered one of the most important topics. In this context, this chapter presents an overview of big data, including their probable advantages and disadvantages, through the analysis of previous studies, using a descriptive approach. In this approach, two directions were noted. The first is *positive*, regarding many characteristics of big data, most importantly the diversity of large-size, and the standard velocity in the analysis, which facilitates the control of costs, time, and human resources; this is to say that the organization's competitive ability is strengthened by allowing appropriate decisions as well. The second is negative, where the inutility of big data has been stated in many studies. In addition, its validity depends on the technological and financial validity of the user information system. Also, from a social point of view, it conducts to the rise of unemployment in sectors that do not need innovation.

## 2.1 INTRODUCTION

Nowadays, a vast amount of data is collected easily, thanks to technological advancements, such as smart devices and applications, the use of credit cards, municipal digital records, etc.

Therefore, big data optimization provides a wide range of advantages, clearly illustrated in terms of its role in the decision-making process and performance enhancement. However, big data is not a magic solution for all problems. This context leads to the following question: *What are the pros and cons of big data?*

This question is treated through two axes; the first one is an overview of big data including, their characteristics and their sources; the second axis is interested in the advantages and disadvantages of big data; finally, a conclusion was conducted.

## 2.2   BIG DATA CONTEXT

Big data generates an important part of our daily life; therefore, understanding this concept is highly important.

### 2.2.1   EMERGENCE OF BIG DATA

1.  **Concept and Classification of Data:** Data is the raw version of information before sorting, arranging, and processing. It is classified into:
    i.    Structured data: organized into tables or databases.
    ii.   Unstructured data: this is the largest portion of data obtained daily as text, images, video, messages, and clicks on websites.
    iii.  Semi-structured data: which is a kind of structured data, but not given in tables or databases [1].
2.  **Big Data Evolution:** The first appearance of the term big data was in the early 2000s and took great importance in technical research centers like Gartner, McKinsey, and IBM.

This context had a big interest in politics like the administration of U.S. President Obama, and the European Commission, where big data is considered as an essential asset to the economy, and society, the same as human, financial, and natural resources.

Many scientific institutions have focused their research on this context, such as the American National Science Foundation, the Canadian Council of Engineering Research and Natural Sciences, the American Institute of Electrical and Electronics Engineers, the European Research and Innovation Programmers, Nature Magazine, Sciences Journal, and the Business and Economy Sector.

The concept of big data takes an important place in the media, such as the New York Times, the Wall-Street Journal, and The Economist.

It is expected that all data will be doubled every two years until 2020; where most of the data will not be produced by humans, but by devices connected to each other via data networks, like sensors, smart devices (direct communication, machine-to-machine, smart cities, and self-driving cars) but so far, only a fraction of the value of the data produced through the use of (data analytics) has been discovered. By 2020, it is estimated that 33% of all data will contain information that can be of value when analyzed [2].

### 2.2.2 DEFINITION AND PARTS

1. **Definition:** Big data is defined as: "Stock of information characterized by volume, velocity, and variety, requiring innovative treatment methods different from complex processing to allow the users to improve their vision, thus good decision making."

   As defined by the International Standards Organization (ISO), "a set of data with many characteristics such as size, velocity, variation, and, validity"; which cannot be effectively processed using traditional techniques for ideal exploitation [3].

   Thus this kind of data cannot be stored or treated using traditional databases because of their large size, multiple sources, diversity, and rapid change.

   Big data represents a stage in the development of information and communication systems to meet the requirements of the control of fast data flow; in fact, it is a real, current, and large-size event, with many characteristics including [4]:

   - **Volume:** Referring to the amount of data generated where the value is, determined by the size.
   - **Variety:** Data exist in two categories, where it can be organized and structured, which represents the smallest portion; it can also be, unstructured which is the biggest portion, or a mixture of the two categories called semi-structured data.
   - **Velocity:** The frequency of data occurs as well as the processing of data within a small period of time.
   - **Variation:** Refers to the inconsistency of data, which can affect the processing efficiency.
   - **Validity:** Related to the quality of the data obtained, which requires careful analysis in terms of its utility, sources, and authenticity.

- **Value:** For ideal exploitation of big data, they must be processed by specialists, knowing who to conduct the appropriate analysis; in this case, the data are considered valuable.
- **Variable Value (Variability):** In the sense that the same information or the same data can have different meanings where its value, the value can be determined and appropriately analyzed, based on the context in which it is presented.
- **Visualization:** When using big data, they must be analyzed and exposed in different forms, following their use, and takes several forms such as statistics, figures, geometric shapes, etc.

Regardless of the mentioned characteristics, the analysis of big data aims to treat the problems resulting from these characteristics, and despite the problems, these characteristics are the key that made them very useful and have tremendous applications in various educational, health, and knowledge institutions; as well as industrial, security, and other installations.

2. **Limits of a Big Data System:** In order to organize a service, you must identify the parts that deal with this service and determine the duties and the rights of each part; the big data system consists of several devices interacting with each other, which is briefly explained in Figure 2.1, where the system consists of:

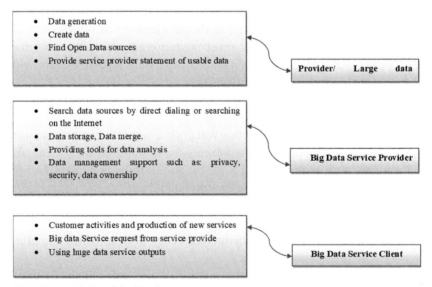

**FIGURE 2.1** Limits of the big data system.
*Source:* Challenge [5].

- **Large Data Provider/Service Provider:** Providing data from different sources to the service provider and includes the activities of data providers.
- **Large Data Service Provider:** The service provider analyzes the big data and provides the necessary infrastructure, and includes the activities of the service provider.
- **Big Data Client:** Which is the final user of a big data system or a system that uses the results or services provided by the big data service provider and the customer can produce new services or knowledge, depending on the results of big data analysis, and include the activities of the client.

## 2.2.3  BIG DATA SOURCES

Nowadays, data are produced automatically and continuously from different digital sources; that can be used in the official statistics with appropriate accuracy and timeliness. The most notable reason for the increase in the data size is that it continues to reproduce much more than before through many devices and sources. And most importantly, most of those statements are not organized, such as tweets on Twitter, videos on YouTube, status updates on Facebook, etc., which means that traditional database management tools and analysis are not useful with these data.

Some big data sources are classified as follows [1]:

- **Program Management Data:** Whether it is a governmental or non-governmental program, such as electronic medical records, hospital visits, insurance records, bank records, and food banks.
- **Commercial Data:** Resulting from the transactions, such as credit cards and transactions on the internet (including mobile devices).
- **Sensor Networks:** Like satellites, roads, and climate sensors.
- **Devices:** Such as tracking data provided by mobile phones and global positioning systems (GPS).
- **Behavior Data:** For example, the number of internet research (on a product, service, or any other type of information).
- **Opinion Related Data:** Such as the comments on social media.

Big data are graphical sources considered as "large-size data, high velocity, and diversity where it requires innovative treatment methods, to be well understood and appropriately used in the decision-making process."

## 2.3    PROS AND CONS OF BIG DATA

Big data offer a competitive advantage to the companies when exploited since the first step started by choosing an appropriate method of collecting, storing, and then analyzing to take advantage of the information provided later; nevertheless, big data have several negative points that we will discuss in this context.

### 2.3.1    BIG DATA PROS

We can cite:
- Exceptional velocity; where many organizations consider that among the advantages of big data is the ability to analyze sensitivity, evaluate offers, thus good decision-making regarding specific goals, added to many other advantages such as increasing productivity, cost reduction, increasing revenue by improving the quality of customer selection, fraud detection, increasing organizational resilience, more creativity, and innovation [6].
- Another benefit for the organizations in the development of new products through the analysis of consumer-related statistics and desires [7].
- Big data analysis adds great value to the organization; by defining comprehensive, accurate, and executive ideas.
- Save of time and money by managing necessary and specific data, thus have a more comprehensive and preferred understanding of the characteristics and needs of individuals and different groups, by including multiple and disparate parts of the data in the analysis process, thereby enhancing the efficiency and accuracy of predictive analyzes.
- Comprehensive analysis of the various organizational and operational processes, thereby increasing the opportunities that strengthen the organization.

### 2.3.2    BIG DATA CONS

Some possible risks of big data, including a lack of clarity in data processing, were explained in an ESMA report (2016). Big data is often deemed unnecessary for some services, considering that they (services) are intended for categories that do not require artificial intelligence. The most significant negative aspect is that some factors such as emotions and privacy are not taken into account; which contributes to an inappropriate interpretation of data [8]:

- Methods of data processing, which is one of the challenges, are that the big data obtained are useless in the "raw" version and must be analyzed using many programs to take information from this huge amount of data and developing plans for data processing. In addition to data quality, where data are unreliable, and incomplete, bad data, in the USA for example, cost 600 billion dollars each year. Moreover, analyzing a great quantity of data needs powerful algorithms and IT tools.
- Security and privacy, where safety is another challenge, given the huge amount of data; this includes user authentication, user-dependent access restriction, log data access history, proper use of data encryption, etc.
- Lack of talent, there are a lot of big data projects in the major organizations, but the presence of a talented team of developers, data scientists, and analysts with enough knowledge in the field represents a challenge [9].
- Prohibitive costs, where it is not enough to have an application to treat this huge amount of data but must be converted to information, and this process is very expensive.
- Big data require a different and changing environment, which makes the process complicated.
- Providing necessary resources and expertise that many companies do not have the strategy should be formulated because the owner big data do not know what to do with them, where most of the companies are suffering from this gap; also, data supervisors should be employed, keep them safe and correct, prevent major problems [10].
- Inappropriate diagnostics; there are people who use the Google search engine (big data) and do not always match the symptoms with the diagnosis.
- The difficulty of formulating public policy perceptions; viewing the information derived from social media is not an effective way because it does not differentiate between what is real and what is false.
- Increasing the burden on the systems; most administrations are not qualified to manage this amount of information and can also surpass the capacity of the workers in the organizations.

- Reducing responsiveness of the users, where many people have no connection with technology.
- The dangers of the occupation of the virtual interaction instead of human interaction: where the misuse of social media has affected direct human communication; likewise, modern technology may push us away from the human relationship that is necessary for human life.

## 2.4　CONCLUSION

We conclude that depending on technical and cognitive growth, the use of big data differs from one nation to another and from one organization to another. Big data strengthen the early warning mechanism for appropriate decision-making and can highlight the real danger component of security and privacy. At the same time, the negative points can be handled by recruiting trained experts to handle the different types of data, thus optimizing benefits.

According to Dean Stoecker (Chairman and CEO at Alteryx), the economic value of data will be the same as oil in the future. The limits of this study are the definitions of big data, the most important characteristics, and sources of big data, as well as an overview of researchers in the field of business management of big data; we did not address the quantitative aspect of the study, this will be the subject of future studies.

In this context, we can suggest the following:

- Developing countries need to be interested in exploiting big data as a new economic power.
- The need to create new jobs in line with real technological development, such as an analyst or expert of big data.

## KEYWORDS

- **big data**
- **big data sources**
- **global positioning systems**
- **information system**
- **program management data**
- **sensor networks**

# REFERENCES

1. Statistics Center, (2013). *General Concepts about Big Data* (p. 5). Abou Dhabi: Guide of methodology and quality.
2. Racheoin, A. M., (2018). The role of big data analysis in rationalizing financial and administrative decision-making in Palestinian universities-case study. *Journal of Economic and Financial Studies, 11*(01), 27.
3. Adviser, (2017). *The Benefits and Risks of Big Data*. From: https://devpr.cleveradviser. com/big-data-financial-technology/ (accessed on 22 October 2020).
4. Abouacha, A., Rafif, T., Asma, N., & Yad, R., (2019). *Big Data with Machine Learning* (p. 2). Damascus University of Information Engineering, Syria: Department of Artificial Intelligence.
5. Challenge, (2019). *The Benefits and Risks of Big Data,* www.gosolis.com (accessed on 22 October 2020).
6. Khaled, A. S., & Abdulah, B. A., (2018). Big data in the libraries of Sultan Qabous University, the impact and the role of manager as an intermediary variable to benefit from in improving services. *Iraqi Information Technology Magazine, 9*(01), 28.
7. Albar, A. M., (2017). *Big Data and Areas of Application* (pp. 2, 4). University Al Malik Abdaziz. Jeddah; Saudi Arabia; Based on Albar.
8. Manager, (2017). *Benefits: The Competitive Advantage of Big Data in Business.* www. newgapps.com. (accessed on 22 October 2020).
9. Ahmed, A. B., (2017). *Big Data has its Characteristics, Opportunities, and Power.* www.alfaisal-scientific.com (accessed on 22 October 2020).
10. Harvey, (2018). *Big Data Pros and Cons*. https://www.datamation.com/big-data/ big-data-pros-and-cons.html (accessed on 22 October 2020).

# CHAPTER 3

# Big Data Uses and the Challenges They Face

NADIA SOUDANI and DJAMILA SADEK

*Faculty of Economics, Business, and Management Sciences, University Center of Tissemsilt, Algeria, E-mails: soudani_mag@hotmail.com (N. Soudani), sadek.djamila.ecom@gmail.com (D. Sadek)*

## ABSTRACT

Big data is considered as one of the most important themes that have been reached in recent years. This is due to its diversity and its benefits if used best. However, its large size has become an obstacle to its best exploitation.

This study aims to identify the big data, types, and characteristics that distinguish them from other data. In addition, this research discusses the most relevant fields in which big data is currently used and how to improve their uses through a strategy that allows dealing with its challenges.

## 3.1 INTRODUCTION

Big data has become widely used and has been used in many sensitive areas such as medical, military, economic, etc. Due to the novelty and constant renewal of this information, the process of use is hampered by a set of obstacles, namely the size and difficulty of storage, the inability to use them fully, etc.

Through this forward, we can pose the following question: *To what extent was big data used?*

1. **The Method Adopted:** The inductive approach is adopted with two analysis tools in data analysis.

2.  **The Importance of the Study:** The importance of the study is reflected in the attempt to research the true meaning of the mega-data and the areas in which it was used, in addition to addressing the obstacles that stand in the way of its use.

The study was divided into the following axes: Section 3.2: An overview of big data; Section 3.3: Big data uses; and Section 3.4: Scientific examples of the uses of big data and the challenges it faces.

## 3.2   AN OVERVIEW OF BIG DATA

Big data has been circulating for a few years because of its importance. This term was not common before, and that is why we must stick to the definition of this term, its several types and characteristics, and identify its different sources.

### 3.2.1   DEFINITION

In 2011, the McKinsey Global Institute introduced a large data set that exceeds the capacity of traditional databases of points: storage, management, and analysis of that data. And in the same vein, the term big data in the field of information technology was launched with a range of packages – very large, complex data that is difficult to handle with traditional database management systems [1].

The United Nations describes big data as data sources of huge sizes, high speeds, and diversity of data, which require new tools and methods to capture, save, manage, and process in an effective manner [1].

Big data is a collection of sets of large and complex data with unique characteristics (such as size, speed, diversity, variation, data health), that cannot be handled efficiently using current and traditional technology to make use of them [2].

These are data that cannot be stored or processed using traditional databases due to their large size, the multiplicity of sources, rapid change, and variety.

### 3.2.2   BIG DATA TYPES AND SOURCES

Data can be categorized into three types [3]:

i.  **Structured:** It is the data that can be stored and processed in a consistent format in the name of structured data. Data stored in rdBMS is one example of structured data. It is easy to process data because it contains a fixed chart. SQL is often used to manage this type of data.

ii. **Semi-Structured:** XML files and JSON documents are examples of this type of data, which does not include the structure of the data model, described in relational databases' management system (DBMS), but has some organizational features, such as tags, to distinguish the semantic elements that make it easier to analyze.

iii. **Unstructured:** Data that has an unknown format and cannot be stored, and cannot be analyzed unless it is converted into a structured format, text files, and multimedia content such as images, audio recordings, and videos are examples of unstructured data. Unstructured data is growing faster than others are, and experts say that 80% of the data in an organization is unstructured.

Therefore, we can categorize some elements or the characteristics that characterize big data in the general framework are called the five V's because they all start with the letter V: Volume, Variety, Velocity, Veracity (reliability), and Value [4]. In the context of big data, the following V's can be provided and taken into account:

- **Volume:** The amount of data we release daily from the content.
- **Variety:** The diversity of these data is between structured and non-structured and semi-structured.
- **Velocity:** How quickly data occurs, for example, the speed at which tweets are posted is different from the speed at which remote sensors can scan climate change [5].
- **Veracity (Reliability):** What is the reliability of the data source, the accuracy, validity, and updating of such data as one in three executives do not trust the data offered to them for decision [2].

## 3.3   BIG DATA USES

This section addresses big data uses and importance.

### 3.3.1   THE IMPORTANCE OF BIG DATA

Big data importance lies in the fact that this phenomenon can:

- Big data can be of great value by making the information transparent and widely usable.

- When organizations transfer all their transactions electronically, they can collect more accurate and detailed performance information and thus display the contrast and improve performance.

- Big data allows customers to be more uniquely segmented, thereby providing more accurately designed products or services.

- Cutting-edge analytics can significantly improve decision-making.

- Big data can be used to improve the development of the next generation of products and services [6].

- Big data offers a competitive advantage for organizations if they are well utilized and analyzed because they provide a deeper understanding of their customers and their requirements and this helps to make decisions within the organization more effectively based on the information extracted from customer databases and thus increase efficiency and profit and reduce waste.

We note through Figure 3.1 that there is a constant increase in interest in big data, especially after 2010, because of the need for information to predict the events, crises, and fluctuations that may occur in the world.

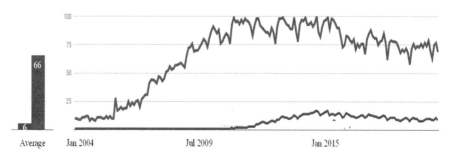

**FIGURE 3.1**   Trends in big data and logical analysis for 2004–2017.
*Source:* Adapted from Ref. [7].

The main reasons for using big data and the benefits they can bring in business are the following summarized in Figure 3.2.

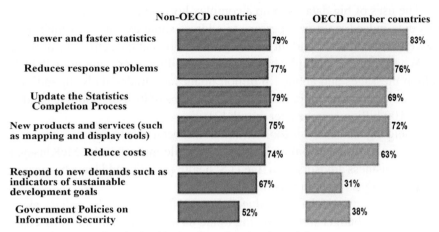

**FIGURE 3.2** Benefits of using big data in performing the activity.
*Source:* Economic and Social Council, United Nations.

### 3.3.2 *WHEN USE BIG DATA?*

Big data is useful for:

- Supporting scientific research in various fields and fields by providing databases at detailed levels of the research communities.
- Encouraging creativity in various economic fields such as using databases to stimulate investment and job creation, in addition to encouraging competitiveness in businesses, which increases the efficiency and quality of services and goods produced.
- Making big data available to improve the level of different services, for example, in the field of health care, for example, using them in research on ways to provide optimal health care and other issues.
- It is an engine of innovation and it opens up new horizons for government and business owners to improve processes and services.
- It helps to explore new areas for growth in all sectors.
- It provides the opportunity for businesses to discover, design, and generate new services and products to meet previously unmet needs.

In addition, there are several factors that contribute to high expectations regarding big data, such as the rapid growth of high-speed broadband, the growing power of cloud computing, access to big data technology in easier and cheaper ways., provides communication between machines.

The uses of big data open up a complex horizon about the actual uses of this data indeed, so it is necessary to address some examples of their uses, as well as the most important challenges that prevent the actual use of them.

WhatsApp, for example, has more than 1 billion users, and more than 42 billion messages and about 1.6 billion images are traded daily. In addition, Facebook deals with more than 50 billion photos of its users. Google handles about 100 billion searches per month [2].

Using big data analysis tools, Walmart has been able to improve its search results for its products online by 10–15%, while in a report by McKinsey, a leading business advisory firm, the U.S. health sector would have been using big data analysis techniques effectively.

Efficiency would have produced over 300 million U.S. dollars as an annual health budget surplus, owing to the 84% reduction in spending costs. According to a previous Gartner Foundation report (2013), 64% of companies invested in the use of digital technologies to deal with big data.

The use of big data does not depend on institutions and businesses but extends to many areas, including energy, education, health, and large scientific projects, most notably the human genome project (the study of the entire genetic material of humans), which contains 25,000 genes, which in turn contain 3 billion base pairs (bp) of chemical bases for DNA [7].

The Great Hydroshock has 150 million sensors that deliver data, 40 million times per second. There are approximately 600 million collisions per second. However, dealing with only less than 0.001% of the sensor current data, the flow of data from all four Collider experiments represents 25 petabytes.

Amazon.com processes millions of back-end operations every day, as well as inquiries from more than half a million third-party vendors. Amazon relies primarily on Linux to handle this huge amount of data, and Amazon has the world's three largest Linux databases with a capacity of 7.8, 18.5, and 24.7 TB.

Windermere Real Estate uses anonymous GPS signals from nearly 100 million drivers to help new home buyers determine their driving times to and from work during different times of the day. Many examples that prove that big data can generate value for businesses which deal with its challenges, including (see Figure 3.3 [8]):

- The volume of big data is constantly increasing;
- Massive and accelerated growth in data quantity;
- Search and random retrieval inside big data;

- Diversity of data;
- Availability of specialized staff in big data analysis;
- Provides expert, automated systems that suit the organization's needs and have good capabilities and flexibility in use and development [8];
- The problem of storing big data;
- Difficult analysis due to too much data;
- Poor security due to the large volume of big data and the inability to keep it safe.
- Lack of talent [3].

**FIGURE 3.3**    The main challenges organizations face in analyzing big data.

## 3.4   CONCLUSION

Big data is the data that have invaded the world in a short time ago, and has become used in several areas, and has become research edited in various other areas, and today studies and research are conducted in order to work on storing them and to find optimal ways to exploit them and try to overcome the difficulties that prevent their application.

The following results were reached:

- Big data is a data that is characterized by its large size, fast flow, and difficulty in storing it.

- Big data is used in different and diverse fields such as medicine, industry, etc.
- The speed and size of big data are one of the most important obstacles to its exploitation and use, in addition to the difficulty of storing it.

Also, the following recommendations are needed:

- Researching ways to store big data;
- Finding ways to educate those who use big data in order to generalize their use in all areas due to their importance;
- Creating programs for big data analysis in order to take advantage of the information it contains.

## KEYWORDS

- **big data**
- **big data aims**
- **big data types**
- **velocity**
- **veracity**
- **volume**

## REFERENCES

1. Statistics Center, Abu Dhabi. General Concepts of Big Data, Methodological, and Quality Guides, Guide No. 13, 4, 4, 5.
2. Al-Bar, A. M. Big Data and Application Areas (pp. 2, 3, 5). King Abdalaziz University.
3. Asma, N., et al. Big Data (p. 2, 9, 10). The University of Damascus.
4. Advanced Electronics Company. The Future of Big Data Analytics in the Middle East, 1.
5. Mohammed, H. A Glimpse of Big Data. tech-wd.com/wd/2013/07/24/what-is-big-data (accessed on 22 October 2020).
6. MCIT, (2018). Nonprofit Technology Index Solutions, Big Data, and its Impact on Improving Nonprofit Sector Performance (p. 10).
7. Aboubaker, S. A., (2017). Big Data, Characteristics, Opportunities, and Strength. https://www.alfaisal-scientific.com/?p=2093 (accessed on 24 November 2020).
8. Ali, B. D., (2019). Big Data and Decision Making at King Saud University. http://www.qscience.com/doi/full/10.5339/jist.2018.15 (accessed on 24 November 2020).

# CHAPTER 4

# Twitter's Big Data Analysis Using RStudio

HOUSSAME EDDINE BALOULI[1] and LAZHAR CHINE[2]

*[1]National High School of Statistics and Applied Economics (ENSSEA), Koléa, Algeria, E-mail: balouli.houssame.eddine@gmail.com*

*[2]Associate Professor, Boumerdes University, Algeria, E-mail: l.chine@univ-boumerdes.dz*

## ABSTRACT

This work is an application of tweets analytics using the R programming language [1] and its interface Rstudio. In this work, we have used many packages such as "twitteR," "tm," and "wordcloud." Two political figures – the US President Donald Trump and the Canadian Prime Minister Justin Trudeau – were chosen because they are among the most influential leaders in the last four years. This research aims to reach a big collection of tweets that talk about each one. After that, data obtained through different steps are revised. Finally, we created a cloud of words (which is a visual representation of words in which the size is proportional to the frequency of that word in a given text) for each one and analyzed them.

## 4.1 INTRODUCTION

Big data is a modern Era that means data sets with several characteristics: high volume, high velocity, and high value. It comes from many areas in our life, such as social media, public health, industry, and agriculture. It is essential to apply some advanced statistical methods in technologies that allow us to extract, store, analyze, and visualize the information [2].

### 4.1.1   BIG DATA ANALYTICS PROCESS

Generally, many steps characterize the big data analytics process: as a first step by identifying the business problem (the subject) to be solved. Next, all sources of data (input) need to be collected; it is an essential step. The next action is data cleaning to make it ready for the analysis. A model, picture, equation, or other type of results will be constructed in the analytics step. Finally, we interpret the final output in order to improve the quality of the decision-making process and clarify the future [3] (Figure 4.1).

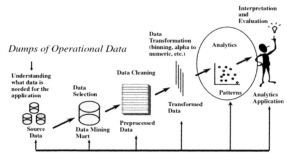

**FIGURE 4.1**   Big data analytics process.
*Source:* Team [1].

### 4.1.2   SOCIAL MEDIA ANALYTICS (SMA)

SMA refers to collecting, cleaning, and treating structured, unstructured, qualitative, or quantitative (quantifiable or not) data loaded from social media websites. Moreover, social media is a broad term encompassing a variety of online platforms that allow subscribers to exchange content, sentiment, information, and others.

Twitter-by its 140 characters rule-is one of the most popular social media websites (after Facebook) with more than 500 million tweets per day. Twitter is growing rapidly since its creation before 13 years. An advantage of Twitter is that all tweets are showed in real-time in which the information can reach a large number of subscribers in a very short moment [4].

### 4.1.3   TWEETS ANALYTICS PROCESS

Tweets analytics process can be described as follow:

i.   **Creating a Connection to the Twitter Server:** By creating new apps on the development website (http://developer.twitter.com/). The server admin will ask you many questions about the nature of your project and the followed approach. The objective is to have access to tweets.

ii. **Select the Data Type:** Select one or many keywords that you need to study and analyze.

iii. **Extract the Data:** Using the package "searchTwitter" or other packages. You can choose any number of tweets you need and can specify the language, the date, and other parameters. All these options are related to the nature of the keywords you search for.

iv. **Clean the Data:** Using the "tm" (text mining) package, we extract only texts from the loaded data. After that, we clean the texts from numbers, punctuations, and other special characters. The next step eliminates a group of English words such as "did," "do," "must," "they," and others. The list of these words will show in the practical part. At this level, we also eliminate repeated words.

v. **Data Processing:** The data now is cleaned and ready for analysis. Using the "wordcloud" package [5], we create the word cloud of both Trump and Trudeau. Many options are available such as colors and the position of the words in the cloud.

### 4.1.4   WORD CLOUD UTILITY

Word cloud is a powerful communication tool. It is very easy to understand and share. The word clouds summarize any topic well. Word cloud is an efficient method for text analysis. It adds simplicity and clarity. The most used keywords appear better in a word cloud presentation. Word cloud is visually more explicable than a data table filled with texts. However, *who uses Word Cloud?*

Word cloud has several uses:

- **Laboratories:** For the presentation of both quantitative and qualitative data.
- **Marketing Campaigns:** To highlight customer needs and identify dissatisfaction.
- **Education:** To support essential topics.
- Politicians and journalists.
- **Social Network:** To collect, analyze, and share the user's sentiments.

## 4.2   EXPERIMENTAL METHODS AND MATERIALS

### 4.2.1   LOADING LIBRARIES (PACKAGES)

We used many packages in our study: "twitteR" [6] to connect to the Twitter website; "tm" [7] to the text mining phase; and "wordcloud" to the presentation.

**Library (twitteR)**

**Library (tm)**

**Library (wordcloud)**

### 4.2.2   CREATING A CONNECTION

The next step is to connect to the Twitter server. For this, we must have authentication keys that can be obtained by registering on the development website (https://developer.twitter.com/). The process is not very complex.

```
consumer_key <- "aSRrP8oQSc29YyvdglqflvogH"
consumer_secret <- "xaA9ihDYXiGWkECOxPC45S6VRzlcnNR29rZWchORLGWqvDgPVw"
access_token <- "1013074431550291968-jxjLtzaELHQB0xqQIrBTkzf2EOsNAg"
access_secret <- "LDEzIC5kw1JwpZK39nsH5gBapE5a93gAVFn7du45zEHKX"
```

Once obtained, we specify the string consumer_key, consumer_secret, access_token, and access_secret in the sutup_twitter_oauth command:

```
setup_twitter_auth (consumer_keyonsumer_secret, access_token, access_secret)
[1] "Using direct authentication"
```

The message "Using direct authentication" should appear in the console, indicating that the operation is running smoothly.

### 4.2.3  EXTRACTION OF TWEETS

The search Twitter function is used to load tweets online. In our study, we specify two keywords: @realDonaldTrump and @JustinTrudeau.

```
tweets_Trump <- searchTwitter ("@realDonaldTrump", n = 5000, since = '2017-01-01')
tweets_Trudeau <- searchTwitter ("@JustinTrudeau", n= 5000, since = '2017-01-01')
```

We limit the number of extracted tweets to n = 5000, and since 01/01/2017 for the date. We are interested in the English language for the tweets. The date of the extraction was 25/01/2019 at 16:00 UC.

### 4.2.4  THE STRUCTURE OF THE OUTPUT

Using the "str" base function of R, we confirm that the output is a list. It means that our output contains characters, numbers, and other types of data.

```
Str(tweets_Trump)
[1] List of 5000
```

### 4.2.5  FIRST TWEETS

We can show any tweet we need, and we can know many things about it such as the date of publication, by who, its IP address, and other information.

```
print(tweets_Trump[[1]])

[1] "keikomeff: @realDonaldTrump Lock him up! You□re next! Mueller is coming!"

print(tweets_Trudeau[[1]])

[1] "Cheryl_Wildlife: RT @Pam_Palmater: Good grief. INAC ALWAYS blames some inanimate object li
ke a law or policy for why horrific things are done to First Natio□"
```

### 4.2.6   CLEANING THE OUTPUT

The first step is the extraction of the text-only using the "sapply" function.

```
Trump_text <- sapply(tweets_Trump, function(x) x$getText())
Trudeau_text <- sapply(tweets_Trudeau, function(x) x$getText())
```

Then, we create a corpus using the "corpus" function.

```
Trump_corpus <- Corpus(VectorSource(Trump_text))
Trudeau_corpus <- Corpus(VectorSource(Trudeau_text))
```

After that, we clean the corpus from numbers, punctuation, special characters, spaces, and a group of English words such as "they," "you," and "must." Now, the text is ready to analyze by the construction of the word cloud.

## 4.3   RESULTS AND DISCUSSION

The first result is the Trump word cloud created using the "wordcloud" function. We limit the showed words to 100.

```
Trump <- wordcloud(Trump_clean, random.order = F, max.words = 100, scale = c(3, 0.5), colors = rainb
ow(60))
```

The word cloud of Trump is given in Figure 4.2.

Through the cloud of words, we notice that there are many words related to different contexts: political, economic, and others. These words are government, democrats, Roger Mueller (The Special Investigator about the possibility of Russian intervention in the elections), border (Mexican border), Maduro (Venezuelan President), FBI, and Nancy Beloucy (President of the US House of Representatives). These could show us all the problems that the American president is suffering and his intervention, even in cases outside the United States (Venezuela, for example).

**FIGURE 4.2**   Trump word cloud.

The word cloud of Trump allows us to discover many things about his personality, his thinking, his vision about many political, social, and economic events, and his interaction with the outside world.

The second-word cloud is about Canadian Prime Minister Justin Trudeau. We use the same function as the precedent Trump word cloud (Figure 4.3).

**FIGURE 4.3**   Trudeau word cloud.

Through this word cloud, we cannot know a lot about Trudeau using this simple analysis. There are no significant words except Canada or Canadian. Although we got more than 5,000 tweets about the Canadian Prime Minister, we could not analyze his personality or know much about him.

## 4.4   CONCLUSION, LIMITATION, AND FUTURE RESEARCH

Unlike some previous studies published in non-peer reviews that have shown how to extract only tweets, we compared extracted tweets in a specific area (politics), and we explain how they are used to extract important information.

Twitter's big data is a treasure that we need to conserve, discover, and analyze. Many personal data, statistics, emotions, visions, reports, plans, strategies, and other types of data are available to analyze.

The study of tweets is a strong focus of the analysis of social networks because Twitter has become an important factor in communication. This example shows that it is easy to initiate the first analysis from data extracted directly online. The data preparation phase is becoming as important as ever.

On the other hand, the programming language R with its interface Rstudio allows us to use it as a powerful tool to extract big data online and clean it to be ready for use and study. It allows us to build many powerful plots and graphs that help managers, researchers, governments, and other actors in the decision making process.

Regarding the limits of the study, the word cloud is not enough to extract information about the studied keywords because the process does not always work well. We need to improve the study by advanced techniques such as sentiment analysis that is the subject of our next work.

## KEYWORDS

- **big data**
- **programming language R**
- **Rstudio**
- **Trudeau word cloud**
- **Twitter**
- **word cloud**

## REFERENCES

1. Team, R. C. R., (2018). *A Language and Environment for Statistical Computing.* Retrieved on CRAN: URL: https://www.R-project.org/ (accessed on 22 October 2020).
2. B, B., (2014). *Analytics in a Big Data World: The Essential Guide to Data Science and its Applications.* John Wiley & Sons.
3. Gandomi, A. H., (2015). Beyond the hype: Big data concepts, methods, and analytics. *International Journal of Information Management,* 137–144.
4. Zhao, Y., (2013). Analyzing Twitter data with text mining and social network analysis. *The 11ᵗʰ Australasian Data Mining and Analytics Conference* (AusDM 2013).
5. Fellows, I., (2018). *Word Cloud: Word Clouds.* Retrieved on CRAN: https://CRAN.R-project.org/package=wordcloud (accessed on 22 October 2020).
6. Gentry, J., (2015). *TwitteR: R Based Twitter Client.* Retrieved on CRAN. https://cran.r-project.org/web/packages/twitteR/index.html (accessed on 22 October 2020).
7. Hornik, I. F., (2018). *Tm: Text Mining Package.* Retrieved from CRAN: https://CRAN.R-project.org/package=tm (accessed on 22 October 2020).

## CHAPTER 5

# Big Data for Business Growth in Small and Medium Enterprises (SMEs)

RABIA AHMED BENYAHIA

*University of Djilali Bounaama, Khemis Miliana, Algeria,*
*E-mail: rabiebenyahia33@yahoo.com*

## ABSTRACT

Nowadays, big data are considered an extremely valued asset, and its use is foremost important. Small and medium enterprises (SMEs)[1] are discovering ways to use this data for their business growth. This study tries to focus on a literature review of academic research, providing an overview of how SMEs can use big data analytics. It's found that SMEs need mechanisms, tools, and methods to acquire Data analytics skills than exploit the potential of big data.

## 5.1 INTRODUCTION

The amount of digital data continues to increase. This proliferation of data is due to the increasing digitization of all areas of the web and the economy. It is in this context that big data was born; by merging various data sources, structured or unstructured, such as using the Internet on mobility, social networks, geolocation, the cloud, measuring vital data, and media streaming.

The volume of digital data has required thinking of new ways of analyzing, sharing, capture, and storage of data. Thus was born the "big data." It is a concept for storing an unspeakable amount of information on a numerical basis.

---

[1]SMEs: Economic literature contains significant differences in the definition of small and medium businesses. Statistical agencies, international organizations, and governments of independent countries emerge with different definitions and categorizations for businesses that do not reflect their differences.

According to the Association for Computing Machinery (ACM) digital library archive in scientific articles about the technological challenges of visualizing "large data sets," this name appeared in October 1997.

In the early 2000s, the concept of "big data" gained momentum. The industry analyst Doug Laney defined big data as the three V's:

1. **Volume:** Refers to the amount of data from different sources, for example, business transactions, social media, and information from account to account.
2. **Velocity:** Refers to the speed of data which must be dealt with it, for example, Radio-frequency identification (RFID tags[2]).
3. **Variety:** Refers to the different types of data, for example, numeric data, e-mail, video, audio, financial transactions, and stock ticker data.

Using big data helps many companies to increase their revenues. On the other hand, the lack of expertise in big data makes many SMEs fail to use it and benefit from it.

This chapter is focused on answering this question: How SME's can use big data analytics to their advantage? In addition, to attempt this, we organized the chapter as follows: Section 5.2: Big data analytics. Section 5.3: Infrastructure for big data analytics. Section 5.4: The advantages of using big data analytics by SMEs. Section 5.5: Factors condition the poor adoption of business and big data analytics by SMEs.

## 5.2   BIG DATA ANALYTICS

The spread of mobile devices, social media, and platforms, including YouTube, has contributed to reaching an impressive volume of information (i.e., in 2012, Facebook users posted 700 status updates per second worldwide).

In 2011, about 4 billion mobile-phone users were identified worldwide; about 12% of them using Smartphones shaving the capability of turning themselves into data-streams. Meanwhile, the video platform, YouTube, received 24 h of video every 60s [2].

---

[2]Radio-frequency identification (RFID) uses electromagnetic fields to automatically identify and track tags attached to objects, people, or animals to relay identifying information to an electronic reader by means of radio waves [1].

The literature identifies 'big data' as the next management revolution and that 'big data' is bringing a big revolution in science and technology. So, the 'big data' call for a radical change to business competition. This phenomenon is capable of unlocking an organization's business value and facilitating firms to tackle key of their business challenges [2].

The term big data describes unstructured data that need sophisticated methods and technologies for data analytics. An International Data Corporation (IDC) report forecasts that big data analytics incomes will be more than $203 billion in 2020, as mentioned in Figure 5.1.

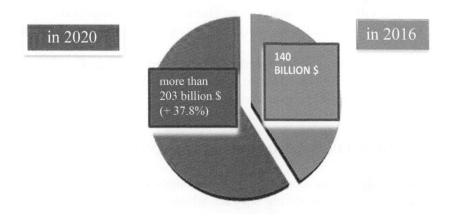

**FIGURE 5.1**     Global big data market.
*Source:* Double-digit growth forecast for the worldwide big data and business analytics market through 2020 led by banking and manufacturing investments, according to IDC. Available at:     https://www.businesswire.com/news/home/20161003005030/en/Double-Digit-Growth-Forecast-for-the-Worldwide-Big-Data-and-Business-Analytics-Market-Through-2020-Led-by-Banking-and-Manufacturing-Investments-According-to-IDC [3] (Significantly modified).

Big data present the opportunity to change business models, government, human resources, and science. Many industries have tried to develop their ability to benefit from the data [4]:

- By processing billions of transactions, Credit card companies can identify fraudulent purchases with a high degree of accuracy.
- Mobile phone companies analyze subscribers' calling patterns to determine, for example, whether a caller's frequent contacts are on a rival network.

- For companies such as Linked In and Facebook, data itself is their primary product. These companies are derived from the data they gather their intrinsic value.

The features of big data can be characterized by 5V's; Volume, high Velocity, high Variety, low Veracity, and high Value [5]:

- **Volume:** A large amount of data that need big storage (Exabyte, Zettabyte).
- **Velocity:** Frequency or speed of data generation.
- **Variety:** Refers to the different types of data from different sources.
- **Veracity:** Big data analysis can give us a reliable prediction.
- **Value:** The benefits through extraction and transformation of big data.

In general, 'big data' is defined as a new approach to manage data and analyze 5V's to create an actionable vision for developing performance and establishing sustainable competitive advantages [2].

## 5.3   INFRASTRUCTURE FOR BIG DATA ANALYTICS

In order to support the aforementioned challenges, there is a standard reference architecture for analyzing big data, which is characterized by a multi-layered structure, as depicted in Figure 5.2.

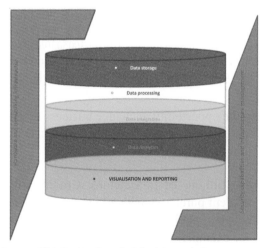

**FIGURE 5.2**   Reference IT infrastructure for big data analytics.
*Source:* Adapted (significantly modified) from Shirley Coleman et al. [6].

The architecture exhibits seven main layers:

- **Hardware and Networking Infrastructure:** Consisting of the physically connected devices providing the core computational power and memory requirements.
- **Data Storage:** Representing the software components to store and manage large data repositories. We can further distinguish two different sublayers:
  - o **File Systems:** Distributed file systems providing storage, fault tolerance, scalability, reliability, and availability. Core technologies in this layer are, for example, the Google file system and the Hadoop system.
  - o **Datastores:** These are the evolution of traditional application databases, which guarantee high-performance distributed access and querying of data into heterogeneous formats. NoSQL databases feature flexible modes, support for simple and easy copy, simple application programming interfaces, eventual consistency, and support of large volumes of data.
- **Data Processing:** Representing the core technologies and frameworks for streaming, interactive, real-time, batch, and iterative data processing. These frameworks work on top of the data storage infrastructure and provide support to more complex data analytics and integration components by enabling scalable primitives for the access and management of data.
- **Data Integration:** Representing the supporting technologies for data ingestion, extraction, transformation, loading, and metadata management.
- **Data Analytics:** Refers to the analytical tools, which help us to explore, predict, and analyze data and learning. Besides that, this layer includes libraries and systems supporting out-of-the-box implementations (such as R, MATLAB, Mahout, MLLib).
- **Visualization and Reporting:** For outputting results in support of the interpretation phase.

Finally, the orthogonal layer of infrastructure management-the operational frameworks for security, benchmarking, and performance optimization to manage workloads, resource scheduling, and management and activity coordination.

## 5.4   THE ADVANTAGES IN USING DATA ANALYTICS BY SME'S

SMEs can derive value from voluminous data by utilizing and developing partnerships with big data technologies, be they in supply chain management, logistics, customer, and business insight, etc. The taking up of big data in SMEs can be fruitful in tackling key challenges of business.

As a matter of fact, the possibilities of big data are not fully explored by people and companies, but individuals are continuously finding big data terms on Google to discover its benefits (Figure 5.3).

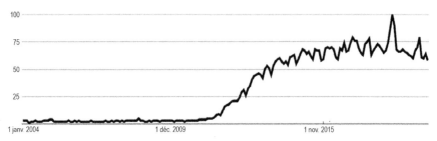

**FIGURE 5.3**    Google trends for big data.
*Source:* Muhammad Iqbal et al. [7]. https://trends.google.com/trends/explore?date=all&q=%2Fm%2F0bs2j8q.

Large enterprises are consistently buying big data and machine learning startups. However, SMEs can also employ big data to better understand their clients, enter into a new market, and cut down unnecessary costs of business, all in real-time. However, the investment in big data is the real challenge to SMEs, so: how can SMEs take advantage of big data?

Here are some of the ways in which small companies can benefit from using big data [8]:

- **It Helps Solving Problem:** SMEs can use big data for problem-solving, Collecting data about processes that allow gaining a new perspective of management. Big data generally gives clarity to small businesses about what is right, what is wrong, and what are the things they can do to improve. However, sometimes things are not so apparent, and companies need to do an in-depth analysis and take things into account before getting any kind of insight. Whatever the case is, one thing is for sure, and that is, big data very important for small businesses.

- **It Helps in Finding Gaps in the Market:** Small companies can become medium-sized companies with the help of big data. In addition to this, it helps large-scale organizations by uncovering new opportunities for them in operations and assesses the risks.
- **It Tends to Add More Efficiency:** Organizations need to use big data in the right manner to reduce their cost. SMEs have reasons to use big data to take advantage of its efficiency benefits.
- **It Gives Added Flexibility:** This phenomenon opens up new opportunities for SMEs to gain a competitive advantage. Since there is so much demand for this, service providers do understand that upfront investments can't be afforded by small-scale operators. Using big data when the terms are flexible works in favor of small companies, as they can use it in whatever way it suits them.

## 5.5   FACTORS CONDITION THE ADOPTION OF BUSINESS AND BIG DATA ANALYTICS BY SMES

Various factors condition the poor adoption of business and big data analytics by SMEs. Among these, we have identified the following ones as being more pervasive and relevant [9]:

1. **Lack of Understanding:** The e-skills UK8 survey highlights an extremely low understanding of big data analytics by SME representatives. It is clear that SMEs will not step into a domain which they seemingly do not understand. Most SMEs are unsure whether their data has at least one of the big data dimensions, and therefore, whether investing in data science is going to bring the benefits claimed by big data enthusiasts.
2. **The Dominance of Domain Specialists:** Operating in a highly specialized field is a particular strength of many SMEs. The major part of the staff is domain specialists. General management functions are poorly covered. Hence, there is reduced awareness of new business trends and opportunities, such as business and big data analytics.
3. **Cultural Barriers and Intrinsic Conservatism:** Domain-specialized SMEs often used to have little interest and confidence in management trends. This attitude can lead them to classify business and big data analytics as management hype rather than as a perspective opportunity. Another issue is infrastructure. Few organizations

set themselves up purely as data industries and thus, when they decide to commit to a big data project, they realize that their data is not accessible or in the format that is desired; when a data specialist comes into the organization. It is not a simple task to extract value from the data.

4. **Shortage of In-House Data Analytics Expertise:** Most SMEs have few or no in-house data-analytic expertise to approach advanced big data analytics. Various factors hamper the creation of adequate in-house expertise: (i) high set-up costs relative to uncertainty in future returns from data analytics; (ii) lack of management expertise to design, establish, and monitor a data analytic unit; (iii) shortage of qualified workers, excessive staff costs.

5. **Bottlenecks in the Labor Market:** There is an increased shortage of qualified data analysts in the labor market. Predict that in 2018, the e-skills UK8 study expects that the demand for big data specialists will increase from 2013 until 2018 by 243. The e-skills UK8 survey reveals that 57% of all recruiters experienced difficulties in filling big data analysis positions.

6. **Lack of Business Cases:** The availability of exemplary case studies and success stories is an important factor for the successful propagation of innovation in business and industry.

7. **Shortage of Useful and Affordable Consulting and Business Analytics Services:** A major part of consulting services used by SMEs concerns the operational level, for example, accounting or hardware-related and software-related IT issues. Management and business analytic consulting are less considered by SMEs. One major reason is that the large consulting companies dominate the management-consulting sector whose business practices are not in line with SMEs' needs and financial capacities.

8. **Non-Transparent Software Market:** Plenty of business analytics software solutions exist on the market. For users with little or no expertise, it is hard to select a product with a good price-performance ratio and to separate the wheat from the chaff. The existing comparison and evaluation platforms are strongly vendor biased. Independent evaluations and selection schemes are hard to find.

9. **Lack of Intuitive Software:** The present market offer in business and big data analytics is split into two extreme parts: potentially useful but highly complex solutions requiring the expertise of knowledgeable data scientists and some simple but less-effective

implementations. Solutions with both an intuitive user interface and a strong analytical potential are rare. IBM's Watson Analytics is one of the few exceptions. Market analysts emphasize the need for predictive analytics software with intuitive user interfaces and a shorter learning curve.

10. **Lack of Management and Organizational Models:** To make business analytics an economic success, a company needs an appropriate management concept and organizational structure. Management challenges in business analytics and big data analytics have been addressed in the literature. Organizational issues have been considered particularly in the context of maturity models. However, the discussion hitherto concentrates exclusively on the requirements of large companies. For instance, issues like leadership, allocation to departments, horizontal, and vertical relationships, and centralized versus distributed functions have little or no relevance for SMEs. The suggested maturity models have rather an assessment purpose than the purpose of providing constructive, detailed advice on how to build up and maintain business analytics in a company.

11. **Concerns on Data Security:** Data security concerns are a key obstacle in the SMEs' path to big data analytics. In an international survey among 82 companies, 22 about 50% of the respondents identified data protection and data security concerns as a barrier for big data analytics. The data security issue is more serious for SMEs than for larger companies. In general, the conditions of and expertise in IT security are at a lower level in SMEs than in bigger companies. An important security gap at SMEs is the use of outdated and unsupported database management systems. Microsoft Windows Server 2003, a major platform used by SMEs, is a notorious example of the aforementioned situation because Microsoft is ending regular support for that software in mid-2015. Consequently, SMEs are more exposed to data breaches and are more vulnerable to intrusion and cyber-attacks. According to recent surveys, 80% of SME's cyber-attacks resulted in PCI (payment card industry) compliance fines, 62% of breaches were targeted at SMEs, 60% of this close within six months of an attack, and 40% of all targeted cyber-attacks was directed to SMEs. The big data environment implies further challenges. Large volumes of data are transmitted through multi-user and multi-owner channels, particularly in supply chains. Being unable to create an in-house data analytic environment, SMEs will resort

to outsourcing analytics services, with a further loss of control over data. Security concerns are particularly serious with respect to cloud services.

12. **Data Privacy and Data Protection:** Customer data processing and analysis have to obey legal constraints on data privacy and protection. In 2012, the European Commission initiated an extensive reform of the data protection rules in the EU that should lead to a single law, the General Data Protection Regulation. The reform should be completed by 2015.

    The present EU data protection regulations and their implications are considerably intricate and not easily accessible for judicially untrained persons. The Handbook on European data protection law 27 has over 200 pages. Many SMEs cannot afford the expert lawyer support needed to understand all the requirements of the legislation.

13. **Different Venture Concept:** The business model for SMEs is often built around specific market opportunities or the existence of differentiating skills and strategic resources that make them competitive in the local or global market. This focused venture perspective creates the idea that business is only dependent on the way they excel in such dimensions, eventually overlooking other resources at their disposal, as well as new opportunities to improve and diversify their activity.

14. **Financial Barriers:** Numerous studies have identified financial barriers as the main issue for SME growth, for instance. SMEs have less access to debt finance than larger companies, particularly because of imperfect information between banks and SMEs. Limited financial resources cause SMEs to be very careful about new investments beyond their specific business scope.

## 5.6   CONCLUSION

Big data analytics can empower SMEs to notice new changes in their units by analyzing data and making correlations between different objects, but this requires tools and methods outside their small structures.

SME's need a mechanism to get started in order to find the best combinations of organizational models as a method of acquiring data analytics skills. The implementation of big data can enable SME's to embark on a cultural change than exploit the potential of big data.

Regardless of the findings of this study, there are several limitations:

➢ First, this study needs to focus on qualitative and quantitative approaches;

➢ Second, this research does not use an empirical study or case studies which involved successful examples of SME's in big data analytics; and

➢ Third, this study does not consider the difference in the potential of big data analytics between small enterprises and medium enterprises.

## KEYWORDS

- **big data**
- **big data analytics**
- **business growth**
- **data security**
- **intuitive software**
- **small and medium enterprises**

## REFERENCES

1. Dictionary, C. E., (2012). *RFID*. Retrieved on: https://www.dictionary.com/browse/rfid (accessed on 22 October 2020).
2. Wamba, S. F., et al., (2015). How 'big data' can make a big impact: Findings from a systematic review and a longitudinal case study. *International Journal of Production Economics, 165*, 234–235.
3. Double-Digit Growth Forecast for the Worldwide Big Data and Business Analytics Market Through (2020). *Led by Banking and Manufacturing Investments, According to IDC*. Available at: https://www.businesswire.com/news/home/20161003005030/en/Double-Digit-Growth-Forecast-for-the-Worldwide-Big-Data-and-Business-Analytics-Market-Through-2020-Led-by-Banking-and-Manufacturing-Investments-According-to-IDC (accessed on 22 October 2020).
4. EMC Education Services, (2015). *Data Science and Big Data Analytics: Discovering, Analyzing, Visualizing, and Presenting Data* (p. 2). Published by John Wiley & Sons, Inc.
5. White, M., (2012). Digital workplaces: Vision and reality. *Bus. Inf. Rev., 29*(4), 211.
6. Coleman, S., et al., (2016). *How Can SMEs Benefit from Big Data? Challenges and a Path Forward*. Published online (www.wileyonlinelibrary.com). doi: 10.1002/qre.2008.

7. Iqbal, M., et al., (2018). A study of big data for business growth in SMEs: Opportunities and challenges. *International Conference on Computing, Mathematics, and Engineering Technologies.*

8. Rashidi, O., (2018). *How SMEs Can Use Big Data Analytics to Their Advantage.* Consulté le. Retrieved on quick start. https://www.quickstart.com/blog/how-smes-can-use-big-data-analytics-to-their-advantage/ (accessed on 24 November 2020).

9. Coleman, S., et al., (2016). *How Can SMEs Benefit from Big Data? Challenges and a Path Forward.* Published online (wileyonlinelibrary.com). doi: 10.1002/qre.2008.

# PART II

# Big Data and Businesses' Decision-Making Processes

**CHAPTER 6**

# The Role of Big Data in Strategic Decision-Making

AMAL BENSAUTRA,[1] AMEL FASSOULI,[1] and FELLA GHIDA[2]

[1]*PhD Student, Faculty of Economics, Business, and Management Sciences, University of Djilali Bounaama, Khemis Miliana, Algeria, E-mails: bensautra.amal@hotmail.com (A. Bensautra), Amelsabrine2018@gmail.com (A. Fassouli)*

[2]*Senior Lecturer, and Researcher, Faculty of Economics, Business, and Management Sciences, University of Djilali Bounaama, Khemis Miliana, Algeria, E-mail: fghida@yahoo.fr*

## ABSTRACT

This study aimed to determine big data, which required the use of modern and sophisticated technologies to control well of institutions and know the role of this data in strategic decision-making. This is why we used the descriptive and analytic methods. The most notable of what we found is that big data directly affects the strategic decision-making process.

## 6.1 INTRODUCTION

Large and new changes are witnessed in financial globalization, online exchange, e-commerce, and the information revolution. This pushes the countries to race and adopt all developments to emerge and ensure their existence and continuity.

Therefore, today the world lives in a set of changes that participated in supporting the technological domain, which could turn what was fantasy yesterday into a scientific truth today among big data, characterized by its speed and multiple sources and different types that reflect positively on the institutions. Thus, it facilitated choosing the appropriate alternative for the decision-maker.

Applying the big data requires knowing their relation with the decision-making process and their reflection on it to be able to predict their role in the institutions. So the question that can be asked is: *What's the relation between big data and the fact to make strategic decisions?*

The study's importance stems from its exposure to big data and its relation with strategic decision-making by recognizing its concept, development, and role in the strategic decision-making process.

We mainly relied on the analytical approach because it is the most appropriate in such a kind of studies and the appropriate to understand the research problems; i.e., starting with gathering the necessary scientific knowledge about big data and strategic decision, in order to determine its role in the sense that the process of access to result in this study was done in a logical sequence.

In addition, the study targets arriving at a set of purposes most notably:

- Recognizing big data and its development through time;
- Understanding the concept of strategic decision;
- Exploring the role of big data in the decision-making process;
- Clarifying the relationship between big data and the decision-making process by giving scientific examples.

This chapter is organized as follows: the first section dealt with the concept of big data to give a view about this term; we touched on the definition of big data, and we gave a historical sight about big data. The second section dealt with the conceptual framework of strategic decisions. Finally, we dealt with supporting the strategic decision to more clarify and better understand.

## 6.2    THE MEANING OF BIG DATA

We deal in this part with the concept of data and big data and its development through time for giving a view about what is big data.

### 6.2.1    DEFINITIONS OF DATA

Data is a word taken from the German word datum, which means facts in the sense what shows the signification of thing and others, its expresses numbers, words, and symbols or facts and rough statistics which has no rela-tion between each other and not explained and not used yet, which means, it has no real meaning and does not affect the user's behavior.

In another definition: data is an abstract symbol, and the material that can be quantitative can be measured and calculated mathematically or can be qualitative (descriptive) like customs and traditions; it requires certain treatments to convert it to results (information) that can be better utilized.

### 6.2.2   THE CONCEPT OF BIG DATA

There is no clear definition of big data in the concept of data volume not specified by space and time under what we are witnessing from the current acceleration in the development of information and communication tech-nology. What is huge data currently may not be huge in the future, also what is big data for someone, or an Institution may not be considered big for another person or institution.

In 2011, the Mackenzie institute launched a definition of big data as a set of data that is beyond the capacity of the classical databases of points storing, managing, and analyzing.

In the same context, the term big data in information technology has been called a set of very large and complex data packages that are difficult to deal with by classical database systems. On the other hand, the UN classified the big data as a large-scale data sources, high speeds and data diversity, which requires new tools and methods to capture, save, manage, preserve, and process it [1].

It is a set of or groups of big and complex data that has unique character-istics (such as volume, speed, variety, contrast, and data validity); it cannot be processed using current and classical technologies to take advantage of it.

The challenges that come with this kind of data are preserving the priva-cies that come with it [2]. It is simply the data that cannot be stored or dealt with using the old databases because of its large volume and its multiple sources and its fast-changing and diversity.

### 6.2.3   THE BEGINNING OF THE TERM "BIG DATA" SAND ITS DEVELOPMENTS

The first appearance of the term "big data" was at the beginning of 2000, but recent expectations have increased the spread of the term among the first next technical trends; it expected the great importance and the spread of technology research centers such as "Gartner," "Mackenzie," and "IBM Company."

It got the interest of many political circles like Obama's circle; the talk has also spread about the big data in the media such as the "New York Times."

In 2007, humanity could store 290 exabytes and 650 exabytes per second 6.4 zettabytes and could connect 58% of computers for general purposes and the computing power has grown yearly by 28% and the global ability to make twist calls has increased by 23, and the humanity witnessed a spread of unilateral information by 6% through TV channels.

From now until 2020, it is expected that all the data will multiply every two years. Humans didn't produce most of the data, but it will be produced by computers that will connect through the internet. It will include, for example, sensors and smart devices, which connect directly with other devices and from a machine to another.

However, until now, we did not discover but a small part of the value of the data, and in 2020, 30% of all the data will contain information with a big amount.

The digital world includes everything from photos and videos on the phones which are downloaded from YouTube to movies and TV contents and includes more data of classical companies such as bank data and security camera recording in the airports and a big event like the Olympic games. Using the data analysis, it is possible to discover the media use pattern and the relations in scientific and medical data from the separated studies and the cross between medical information and social data as well as the appearance of billions of people in the security camera footage.

In addition, the data that has been looked for the need to be marked with descriptive data to measure it and that includes adding history to visual or information of locating photos or smartphone videos or getting useful information from data stores [3].

## 6.3   THE CONCEPTUAL FRAMEWORK OF STRATEGIC DECISION

Administrative decision-making means careful selection by the administration or the decision-making for a particular action.

### 6.3.1   DEFINITIONS OF STRATEGIC DECISION-MAKING

The decision in the background of the administrative process and its political media in achieve objectives the decision is actually select one from a set of alternatives, exposed to solve a problem or manage a certain work by selecting weak and power points for an alternative as a preliminary to choose the best alternative.

The strategic decision appoint what the organization will be in the future, and its effect will be a global organizational unit such as deciding on the merger of the organization or its size, its multiple definitions. The strategic decision is an exceptional decision. It is manufactured in the current period of high importance in terms of its impact on the organization in the coming time period. It focuses on achieving the organizations' goal by understanding how the decision-making process flows through it, and it requires creative skills to surround the internal and external environment variables.

### 6.3.2   THE LEVEL OF STRATEGIC DECISION

The strategic analysis process takes three levels according to the decision levels with the strategy at the stage of formulating the strategy, but it varies in terms of the component on purpose:

- **Organization-Wide Strategic Analysis:** This level is defined by the conduct of activities that determine the organization's distinctive characteristics from other competing institutions.
- **Strategic Analysis at the Activity Level:** It is the management of strategic business units to enable effective competition in a particular area or market or products and contribute to the organization's objectives.
- **Strategic Analysis at the Functional Level:** Strategic analysis is practiced at the level of the various function of the organization, such as the function of production, finance, marketing, and human resources. According to a strategic perspective, each job is concerned with the exploitation of its resources and the management of its systems, which is important for its communication; for example, the function of production is concerned with the size, quality, and requirement of production [4].

## 6.4   SUPPORTING THE STRATEGIC DECISION USING BIG DATA

The decision-making process is the heart of the administrative process. The success of the institution or government sector depends to a large extent on the ability and efficiency of the administrative leadership to make appropriate management decisions. The decision-making begins with data collection, processing, and debriefing. Many large corporations and government sectors rely on big and complex data analysis. Management and analytics cannot be processed with a single tool on work with classical data processing applications.

It is known that collecting data and information helps to accurately characterize the problem and analyze it to achieve accurate results, so it was necessary to adopt an administrative system that includes the analysis of huge data.

The government sector and large companies use the big data analysis system in order to [5]:

- Improve internal processes such as risk management, customer relationship management, and logistic improvement of old products and services.
- Improvement of new products and services, taking advantage of the information, and presenting the right offers to customers at the right time.

Assume that the ministry of education decided to open or close a school in a particular area. So, the ministry should use data from outside the ministry to know the application's population concentration outside the working hours. For example, the use of telecommunication companies' data on the number of phones connected in the telecommunication tower in a particular region allows them to analyze the users' data to determine the concentration of the target age group in this region. It also compares with the official statistics and links with the data of housing programs and the calculation of citizens who are entered in the official statistics and compares with the data of users joined in the program.

When there is a difference in conflict, it is analyzed to find out the cause. The bottom line is that we live in a world built with permanent terabytes of data. It is growing steadily because of the programs that arise every day and the technological means that data are made available for use anywhere and anytime. In addition to changing social culture, all of this contributes to an unlimited number of data. Many are free to exist in networks and virtual spaces industry unstructured, structured, and half-structured.

Institutions today must seed to create the first culture that accommodates these transformations and also create analytical by linking these data to the relationship that contributes to providing information. The UN, represented by the economic and social council, noted in its reports that the basis for development and statistics for 2017 must be based on bid data and analysis.

## 6.5 BUSINESS EXPERIENCES: EXAMPLES

Amazon.com website analyses millions of background processes daily, as well as inquiries from more than half a million vendors from a third party. Amazon relies mainly on the Linux system to be able to deal with Linux databases with the world, which has a capacity of 24.7 and 18.5 terabytes with this huge amount of data, and Amazon has 73.8.

More than a million transactions per hour, which are imported into the Walmart chain of stores (contains 2.5 terabytes), are as much as 2560 petabytes of data, twice the data in all books in the library of congress, which equals 167 billion photos from the users base. Credit cards are protected. Facebook deals with 2.1 billion online accounts in the world. Windermere company helps 100 million divers to buy a new house and eliminate their driving hours from and work.

## 6.6 CONCLUSION

Under the changes and global directions, institutions endeavored to keep pace with them and adopt all technologies by collecting information and electronically exchange it with institutions or between costumers which reflected positively on its performance which can be said that:

- The degree of decision-making is affected by bout the subject of data and its role in making a strategic decision, so we got to know its conceptual aspect by giving scientific examples from reality.
- Big data is a subject of the times that require institutions to pay attention to it and adopt modern techniques to address them because they are very difficult.
- Strategic decision-making is the fate of institutions and therefore seeks to adopt all methods to make the perfect decision.

- Global organizations are competing for the best use of big data because of their role in strategic decision-making.
- Making the right derision depends on the ability to control big data.

Finally, we suggest the following:

- Pay attention to big data and adopting modern technologies for processing.
- Establish cooperation centers between international institutions for optimal exploitation of big data.
- Strive to establish private educational institutions for how to exploit big data and like the process of strategic decision-making.
- Big data needs competent people to exploit them in the fullest.
- Give importance to big data in Algeria in order to keep up with changes and reach the level of challenging countries on the technological side.
- Pursuit to control big data analysis in order to take advantage of its content.
- Continue researches and studies about this topic because of its importance currently.

## KEYWORDS

- **big data**
- **globalization**
- **petabytes**
- **strategic decision-making**
- **Walmart chain**
- **zettabytes**

## REFERENCES

1. Statistics Center, (2013). *General Concepts of Big Data.* Guide No. 13. www.scad.ae. (accessed on 22 October 2020).
2. Albar, A. M., (2018). *Big Data, and its Application Fields.* Faculty of accounts and information technology, University of Abdul-Aziz. http://www.awforum.org/ (accessed on 22 October 2020).
3. Abou, S. A., (2017). Information and communication technology big data: Its characteristics and opportunities and strength. *Al Faisal Scientific Journal.*

4. Laouissat, D., (2005). *Scientific Administration, and Decision-Making Process, Dar Houma for Production and Distribution*, 6.
5. Saadaoui, Y., (2012). *The role of Administrative Information in Strategic Decision-Making* (Vol. 7, pp. 58, 59). Magazine of economic researches, Saad Dahleb University.
6. Albar, A. M., (2018). *Big Data, and its Application Fields* (p. 71). Faculty of accounts and information technology, University of Abdul-Aziz, Op. Cit.

# CHAPTER 7

# Data Mining and Its Contribution to Decision-Making in Business Organizations

NADIA HAMDI PACHA,[1] FATMA ZOHRA KHEBAZI,[2] and
NACHIDA MAZOUZ[3]

*[1]Lecturer and Researcher, Faculty of Economics, Business, and
Management Sciences, University Ali Lounici-Blida 2, Route d'El Afroun,
Blida, Algeria, E-mail: hamdipacha.n@hotmail.com*

*[2]Lecturer, Faculty of Economics, Business, and Management Sciences,
Khemis Miliana University, Rue Thiniet El Had, Khemis Miliana, Ain Defla,
Algeria, E-mail: fkhebazi@gmail.com*

*[3]Lecturer, Faculty of Economics, Business, and Management Sciences,
University Ali Lounici-Blida 2, Route d'El Afroun, Blida, Algeria*

## ABSTRACT

Data mining (DM) has emerged in response to the need for organizations to find a way to take advantage of the large data stored in their databases and repositories, from which traditional analysis methods are unable to extract useful information from them. This technology, based on intelligent deductive algorithms, allows the conversion of a huge amount of raw data into meaningful information, and into new knowledge, which is commonly used to support decision-making.

Therefore, this chapter seeks to provide a theoretical study on how this technology contributes to support decision-making in business organizations. This is to sensitize organizations and practitioners of the advantages of this technology is adopted and arouse the curiosity of scholars and researchers to enrich the subject by studies and research, especially in economic and administrative sciences.

To reach clear results, the descriptive and analytical approach was adopted in dealing with the subject by reviewing the articles, studies, and reports related to the research problem. These references were used in determining the basic theoretical concepts related to DM and their assumed role in the decision-making process. A practical case was also noted on the application of the technology by looking at the status of Dubai Airports.

This research paper concluded that this smart technology works to search for relationships, trends, and patterns hidden in databases, to be used in building models of prediction, and in exploring the behavior of individuals, and in determining their general trends.

This useful information is used in several areas, such as creating a competitive advantage that the organization focuses on, or improving the performance of its services. The systematic and orderly use of this technology also makes the organization's information system an integrated system that discovers shares, and distributes knowledge.

This allows the provision of accurate and rapid information, which contributes to better decision-making, especially on how to increase profits or reduce costs. This research concluded that Dubai Airports Corporation has been able to benefit from the applications of this technology, not only in supporting the decision-making process but also in creating a competitive advantage, improving services, reducing costs, and raising its overall performance.

## 7.1  INTRODUCTION

DM is a new trend in the success of decision support systems in modern business organizations. The development of this technology is the inevitable result of the rapid development of information and communication technology, digitization, artificial intelligence, and machine learning, which resulted in the provision of a huge amount of data stored in the various databases and stores.

The orientation of organizations towards the knowledge economy has forced them to seek new knowledge to be used in making various decisions that will improve their competitive advantage. In response to these new variables in the business environment, DM technology has been introduced, which has allowed the organization to search within the vast volumes of raw data available to it and extract new, previously unknown, implicit information that could be utilized.

DM allowed the discovery of new knowledge hidden in databases, by identifying relationships that were not discovered by traditional methods of analysis, in addition, to identifying patterns and trends in these data and turning them into information with specific characteristics and evidence, which can be used in the decision-making process.

Accordingly, the problem of this paper is to determine how this technology contributes to the decision-making process in the business organization. In this light, the research attempts to answer the following sub-questions:

- What is data mining? Why is it important?
- How does data mining help to make appropriate decisions that will improve the performance of business organizations?
- What are the applications of data mining in organizations, and specifically in Dubai airports?

Therefore, this paper seeks to achieve the following objectives:

- Introduce data mining technology and its role in providing knowledge to the organization;
- Understand how it can contribute to decision-making;
- Access to technology applications in some fields;
- View practical status by reviewing the status of Dubai airports.

The importance of the research lies in the fact that the field of DM is still a relatively recent topic, both in terms of the literature related to this field and in terms of its practical applications to organizations.

The study of the dimensions of the use and investment of big data in business organizations and in the decision-making process is still the focus of the attention of many scholars, researchers, and decision-makers. Therefore, we considered it important to shed more light on this technique by introducing it and clarifying its effects.

Therefore, we will address this topic by addressing first, the most important elements that will define the technology of DM, which will form the theoretical framework for research. We then review the tools used in this research in order to determine the contribution of this technique to decision-making, in addition to the results achieved. Finally, we discuss these findings in light of the practical situation, indicating the limits of this study and its prospects in conclusion.

## 7.2 THEORETICAL FRAMEWORK FOR RESEARCH

The accumulation of data has become a major problem for all organizations, which requires the existence of new methods of dealing with them. In order to ensure easy access to any required information and as soon as possible, it enables the management to make rational decisions on any aspect of the organization. The accumulated data is general and comprehensive for all parts of the organization and its various activities. DM technology is an appropriate solution to this issue [1].

The term DM appeared in the United States in the mid-nineties, combining statistics and information technology. It relies on the use of modern algorithms in data processing and analysis to reach patterns and relationships in large data sets, which allows the prediction of future behavior. It is a guide to decision-making in all business applications [1].

DM was defined as "extracting useful knowledge from large amounts of data, using machine learning techniques and statistical methods" [2]. It is also defined as "an analysis of data sets that are often great for discovering and summarizing unexpected relationships, in a new and understandable way" [3]. DM is the process by which the organization seeks to find new knowledge hidden in large databases, in which it is difficult to use regular statistical methods for analysis.

Some of the main factors that have led organizations to increase their use of this technique are [4, 5]:

- Increasing the data amount that is produced by organizations, and the limited traditional analytical tools to discover new patterns.
- Expansion of DM applications to many fields such as marketing, banking, insurance, transportation.
- The emergence of new methods of analysis, most notably neural networks, genetic algorithms systems, and induction rules.
- The market competition that leads organizations to make the most of the data, especially with the difficulty of obtaining it and the growing need for rapid analytical results.
- The tremendous development in data storage and processing methods, such as data warehouses and markets data warehouse.
- The emergence of new generations of user-friendly software: Microsoft Windows, Client-Server software.
- Costs of electronic communications reduce, which eases access to databases.

The most important benefit of DM technology is its ability to provide accurate, correct, and fast information, which benefits the organization in many respects [1]:

- Access to data that would not otherwise have been possible;
- Find some kind of deductive data types by examining the records in the data warehouse files;
- Achieve a certain understanding level that contributes to the discovery of knowledge from databases;
- Predict future values resulting from operations and business behavior;
- Discover the hidden links, trends, and new patterns that exist between the vast amounts of business data.

This is done by using two exploration models [6]:

i.  **Prospective or Predictive Models:** A model of the system results from it by described the used data used for exploration and aims to predict the value of certain characteristics, such as the probability of purchase for the customer.
ii. **Descriptive Exploration models:** It produces new information based on the information contained within the user data in the exploration process. In addition, it is divided into clustering models that allow individuals, events, or products to be clustered into; and correlation models that allow the identification of relationships between them.

Data can be explored using several tools; the most important of them are [1, 7]:

➢ **First:** Summarization: it refers to the methods of fragmentation of large data blocks into summary measures, which provide a general description of the variables and their relationships. As summarization methods, we can include averages, totals, and descriptive statistics that include measures of central tendency such as arithmetic mean, median, and mode, and dispersion measures such as standard deviation.
➢ **Second:** Classification: it includes identifying levels within the data to be recognized on the information through correspondence with the proposed levels in advance. It helps to discover new levels and items of information. Classification can be accomplished according to historical statistical methods, such as regression and discriminatory

analysis, or on relatively recent methods such as correlation forces, case-based inference, and neural networks.

➢ **Third:** Prediction: prediction is similar to classify or estimate, except that the data are classified on the basis of predicting the future or their behavior estimation of future value. Traditional tools used in forecasting include regression types and discriminatory analysis. New methods include decision trees, neural networks, and genetic algorithms.

➢ **Fourth:** Clustering analysis or fragmentation: it is a descriptive approach designed to separate homogeneous data from heterogeneous characteristics in a community of individuals. This is done based on information contained in the totals of the variables that describe them, helping develop marketing programs tailored to the customers' own sizes in order to repeat the purchase or transform it into loyal customers. Cluster aggregation methods are assisted by statistical cluster analysis, decision tree-based methods, neural networks, and genetic algorithms.

➢ **Fifth:** Rule analysis: it refers to a set of methods that are used to link buying patterns through cross-sectorial or over time. For example, the analysis of the market basket method, by using the underlying information in the goods purchased by consumers actually, to predict the potential of goods bought them if they are offering special offers or if they are introduced to these goods.

➢ **Sixth:** Regression analysis: it is used to convert the data into an expressive explanatory value that helps in adding value to the prediction and estimation process, such as predicting sales volume or the relationship between variables.

➢ **Seventh:** Sequential analysis: it seeks to find similar models in the qualities that occur during a business succession period.

DM is conducted according to the following stages [1, 8]:

1. **The Stage of Business Understanding:** It is an essential element for the success of the DM process.
2. **Data Understanding Stage:** It is the first stage of the exploration of data. In addition, this by knowing what is the nature of the data, in order to assist designers in the use of algorithms or tools used for specific issues with high accuracy. This leads to maximize the chances of success as well as to raise the effectiveness and efficiency

of the system of knowledge discovery. The required stages of understanding the data are:

- **Data Collection:** It means identifying the data source in the study process.
- **Data Description:** It is the focus on configuring the contents of files or tables.
- **Data Quality and Review:** It is the act of negligence of unnecessary or poor quality data that are not useful in the study in order to ensure accurate data.
- **Exploratory Analysis of Data:** This stage is important because it focuses on developing hypotheses related to the problem under study. The initial analysis of the data is carried out using visualization or direct analysis OLAP.

3. **Data Preparation Stage:** This phase aims to improve the quality of real data to explore and increase the efficiency of the process by reducing the required time for exploration. This stage includes the following steps:

- Selection: The choice of expected variables and sample.
- Construction and transformation variables: To formulate new variables to construct effective models.
- Data integration: storing data sets in multipurpose databases with the aim of consolidating them into a single database.
- Data formatting: Rearranging the data fields by the data-mining model.

4. **Model Building and Validation:** By testing and examining various alternatives to get the best model to solve the problem under study.
5. **Evaluation and Interpretation:** By checking the reliability of the data sets that are growing by the model. Since the results of this data are known, they are compared with the actual results in the stability of the running data set to check the accuracy of the form.
6. **Model Deployment:** It includes the publication and distribution model within the organization to assist the decision-making process.

## 7.3   SEARCH TOOLS AND RESULTS

In order to identify the problem of research, it was reported to reviewing the articles, studies, and research available on the subject. This is to obtain a clear-cut picture of the role of DM technology in providing the organization with the knowledge, how it supports decision-making, as well as its various applications and uses.

A selected practical case was also inferred in order to drop the theoretical research results on the practical reality by reviewing the different applications of this technology in one of the major business organizations in the aviation sector, represented by the Dubai Airports Corporation. After reviewing the websites of many organizations, this case was chosen in view of the tangible positive effects this technology has had on improving airport management and the services it provides to travelers.

By looking at the implications of adopting this technology on organizations, we have found that it leads to a range of positive effects as a result of providing extensive information and new knowledge to help better decision-making.

This technology leads to the creation of a new internal work environment, which makes the organization information system an integrated system [9], which, in addition to their traditional roles, exercise the cognitive role, which provides the organization with the knowledge through [9, 10]:

- **Creating Knowledge:** Research systems on accounting and financial information provide graphics, analytics, and document management tools to those working in the field of knowledge, as well as the provision for sources of information and internal and external knowledge.
- **Discovering and Codifying Knowledge:** DM technology provides the possibility to develop and integrate expertise for the purpose of creating models and relationships in large amounts of data and discovering new knowledge.
- **Sharing Knowledge:** Research systems for accounting and financial information provided by collective cooperation exploration mechanisms help employees to access and work simultaneously on the same document, from different locations, and then coordinate their different activities.
- **Distributing Knowledge:** Research systems for DM techniques and their communication tools can secure documents and other forms of information and distribute them to information and knowledge workers in order to link offices to other business units within and outside the organization.

DM technology also provides accurate and fast information that demonstrates its importance in supporting the decision making process. In addition to providing the following features [9]:

- Assist decision-makers in activating the interdependence between the different divisions and actions of the organization.
- Facilitates the handling of advanced information technologies and helps to measure the effectiveness and productivity of different sub-information systems by providing accurate information.
- Assists in the effective use of available data sources and resources, and in planning and improving the used accounting and banking information systems.
- The information helps to increase knowledge, reduce alternatives, and eliminate the uncertainty.
- Enable the decision-maker to identify the problem and its elements, complete the control of the information and data files necessary for its decisions, and seek to develop accounting information systems at appropriate costs.
- It provides the necessary information to make decisions that contribute to expediting tasks and simplifying procedures.

As a result of these features, technology applications were used in various fields, where the first applications for DM in the field of customer relationship management, by analyzing the behavior of customers in order to maintain their loyalty and propose products according to their wishes.

Distribution organizations or large areas of distribution are the first to use them, and then moved to banks, then insurance institutions, then mobile phone companies, then water and electricity institutions, and more recently, air transport and rail transport institutions, etc. The applications of this technique have also been used in other fields [8]:

- **Marketing:** Artificial neural networks are used in target marketing studies, and they helped the use of customer's approach of the allocation according to demographic realities, like sex, age groups, as well as purchasing.
- **Retail:** These techniques have been used effectively to predict sales, based on several variables such as market variables or customer buying habits.
- **Banks:** The applications of this technology in this field have been used in an excellent way in finding guaranteed prices, future price

forecasts, and stock performance, and in determining the risk of loans and financial fraud.

- **Insurance:** These techniques have been widely used in this area by determining policy prices, and the expected future oscillations, and in identifying counterfeit claims.
- **Operating Management:** Neural networks were used in planning, scheduling, project management, as well as quality management.

In general, business organizations can benefit from DM applications in:

- Writing summary reports on specific categories such as regular customers and credit cards;
- analysis of trade tendency by creating markets with strong or weak growth potential;
- Marketing for certain categories based on common characteristics that are discovered;
- analysis of use by finding a particular pattern for the use of services and goods;
- Compare campaign strategies with each other in order to find the most effective.

As for Dubai Airports, it looked forward to taking advantage of the exploration of large volumes of operational data to lead better customer service through the airport-from waiting times in the safety queue to the toilets [11], especially in light of expectations of high rates of passenger and cargo traffic [12].

It also seeks to increase the capacity of its airports, using existing resources and without resorting to the construction of any additional space, infrastructure, or new runways [13]. To overcome this problem, the corporation considered that the solution lies in the use and analysis of data in airports to increase its efficiency.

Through DM, Dubai Airports Corporation aims to monitor and collect data from all possible sources and optimally use it to increase the capacity of existing airports, improve services, reduce energy consumption, and reduce financial costs [11].

In order to achieve the goal set by DAC for the exploration, the sources of data to be collected from the airports were identified in: flight data, Wi-Fi data, metal detector data, baggage data, various sensors data (door sensors-bathrooms-water taps), etc.) and 3D camera data to fetch queue data [13].

Dubai Airports has entrusted the exploration to a specialized company, Splunk [13], as a result of the vast amount of data collected from previous sources, making the exploration process difficult and expensive and requires considerable resources.

## 7.4   PRESENTATION AND DISCUSSION OF PRACTICAL RESULTS

Dubai Airports operates and develops both Dubai International Airport and Dubai Al Maktoum International Airport [14], currently handling over 10 million tonnes of cargo and over 160 million passengers annually, making The busiest airports in the world [15].

Following the exploration of its operational data, Dubai Airports has been able to achieve its objectives. This process has enabled decision-makers to improve the services provided to customers as a result of providing accurate and fast information. This technique has allowed the identification of situations requiring a particular action and then sending alerts to decision-makers in real-time to take appropriate measures and actions. This allowed the following results [11, 13, 15]:

1. **Reducing the Required Time to Bypass Security Measures:** Dubai Airports has been able to reduce the time required for security measures to five minutes for 95% of its passengers. This was achieved by analyzing metal detector data and 3D camera data. For example, people traveling to cold areas wear heavy shoes that activate the alarm, so an automatic message appears on the screens near the rows informing passengers to take off their shoes to speed up traffic; the results of this data analysis are also shared with the security and safety section to continuously improve services.
2. **Second-Improving Internet Access:** Dubai Airports has been able to offer its passengers the fastest and the best Internet service in the world and this for more than 20,000 passengers at the same time. It was done by monitoring all access in real-time and to identify places where there is a lot of pressure points and resolve the problem directly, in addition to identifying any harmful acts carried out by travelers and stop it.
3. **Increasing the Efficiency of the Baggage Distribution System:** The baggage system at Dubai airports is one of the most complex systems in the world and works to predict the weights of baggage and cargo and delivery to the right destination, and this by placing

a sticker on the baggage allows the generation of information from more than 200 data points, which is conformity with the data related to the operations of the airport and forecasting the expected loads during the coming hours.

4. **Reduce Costs and Improve Cleaning Services and Energy Consumption:** This is to provide all bathrooms and tap sensors, where data from these sensors show the bathrooms and laundries most frequently used or must be verified clean and functioning properly. Based on the information obtained from these data, the necessary number of staff is determined, and clean-up and maintenance operations are carried out in a timely manner, which has allowed energy consumption to be reduced to 20%, which will save about $23 million in 2023.

## 7.5    CONCLUSION

As a result of strong growth in databases and stores and intense competition in the global marketplace, business organizations are increasingly interested in looking for new applications.

In order to make the most of their data, to discover new knowledge that helps to make appropriate decisions, to improve their performance, and increase their competitiveness, a procedural phase was therefore needed to extract specifications and hidden relationships in the vast amounts of business data stored in traditional information systems and to provide new, previously unknown information on which to make new decisions.

This is what is allowed by DM technology, which is part of the discovery of knowledge, which begins with the selection of data of interest and conducts a number of pre-processing operations to be appropriate to the technology of exploration. Then apply that technology and evaluate the results obtained and make the appropriate decision accordingly.

Through this paper, we concluded that DM is a smart technology that allows the analysis, synthesis, and transformation of large data into useful and important information.

This is by creating anonymous relationships, highlighting general trends, and showing patterns in these statements, which were not known to the organization before the exploration. DM leads to internal and external knowledge of the organization by providing sources of information and facilitating access to it by its personnel, which helps to coordinate the various activities and processes.

It also makes the organization's information system an integrated system, which works to discover, share, and distribute knowledge, allowing accurate and rapid information to contribute to better decision-making, especially in terms of how to increase profits or reduce costs.

The best example of how DM applications can be used to support decision-making, create a competitive advantage, and improve the overall performance of the organization is Dubai Airports. With the investment of its data mine, it was able to extract important information from flight data, sensor data, digital cameras, metal detectors, luggage systems, and Wi-Fi data.

In light of this new information, the decision-makers took appropriate procedures and necessary measures to increase the carrying capacity of Dubai airports, improve the services provided, and reduce energy consumption and reduce costs, using only existing resources.

In spite of the important results we have reached through this paper, this research did not satisfy the subject of his right to study. This is because of the multiplicity of its competencies, on the one hand, and the fact that the main objective of this chapter is to introduce this modern technology to non-specialists and to highlight its role to managers and decision-makers as a technology that supports decision-making in the organization, on the other.

This topic, given its importance, opens the door to numerous theoretical and practical studies related to exploring its most important uses in different sectors and fields, whether in business organizations, non-profit organizations, or public sector institutions. For example, the contribution of data exploration to reducing the cost of running government facilities, delivering public services, improving the performance of hospitals and universities, improving the performance of public institutions, or the limits of applying this technology in certain areas.

## KEYWORDS

- **baggage distribution system**
- **data mining**
- **decision making**
- **Dubai Airports Corporation**
- **organization information system**
- **pre-processing operations**

## REFERENCES

1. Sabah, M. M., & Sadiq, H. A. M., (2015). The impact of data mining technology on the development of decision-making: An exploratory study in The Iraqi Ministry of Higher Education and Scientific Research. *Management and Economics Magazines, 38 Years* (Vol. 103, p. 87).
2. Al-Fatlawi, H., (2017). *Forecasting Techniques in Data Mining.* Center for Research and Information Rehabilitation, University of Kufa, [Online]: http://uokufa.edu.iq (accessed on 22 October 2020).
3. David, H., Heikki, M., & Padhraic, S., (2001). *Data Mining* (p. 222). Massachusetts Institute of Technology, USA.
4. Paulraj, P., (2010). *Data Warehousing Fundamentals* (p. 430). John Wiley and Sons, New York, USA.
5. Al-Alaq, B. A., (2005). *Digital Management: Fields and Applications* (pp. 89–91). Emirates Center for Strategic Studies and Research, Abu Dhabi.
6. El-Deeb, K. H. M., (2012). *Modern Methods of Data Analysis (Data Mining), an Analytical Study* (p. 3). [Online]: http://digital.jilwan.com/digital2012/download2012.php?f=jalsa4/4_1.pdf (accessed on 22 October 2020).
7. Berry, J., & Linoff, G. S., (2004). *Data Mining Techniques for Marketing, Sales, and Customer Relationship Management* (2nd edn., p. 10). Wiley Publishing, Inc, Indianapolis.
8. Al-Ali, A. S., Amer, I. Q., & Al-Omari, G., (2006). *Introduction to Knowledge Management* (pp. 100–102). Dar Al-Massirah for Publishing, Distribution and Printing, Amman.
9. Al-Shehadeh, A. R., (2013). The effect of application of data mining techniques on banking operations management. *Damascus Journal of Economic and Legal Sciences, 29*(2), 178.
10. Turban, E., & Leidner, D., (2008). *Information Technology for Management* (6th edn., pp. 183–186). John Wiley & Sons, New Jersey.
11. Scott, C., (2017). *Dubai Airport Turns to Real-Time Data in Drive to be the World's "Biggest and Best Airport."* Computer World, UK, [Online]. https://www.computerworlduk.com/data/npl-relaunches-with-new-focus-on-measurements-supporting-uks-digital-industries-3664417/ (accessed on 22 October 2020).
12. Michael, I., (2014). *Executive Director of Dubai Airports, said: "We Do Not Want to be Bigger, We Want to be the Best, with New Levels of Customer Service."*
13. Splunk, (2018). *Dubai Airports Flies into the Future with Splunk: Case Study.* [Online]: https://www.splunk.com/pdfs/customer-success-stories/dubai-airports-case-study.pdf (accessed on 22 October 2020).
14. *Dubai Airport's Official Website.* http://www.dubaiairports.ae/en/corporate/about-us/history (accessed on 22 October 2020).
15. Al-Khayat, I. *How Dubai Airports Analysis Data to Improve Its Services.* Arabic site, Science [Online]. https://datasciencearabi.com/dubai-airports (accessed on 22 October 2020).

# CHAPTER 8

# The Strategic Role of Big Data Analytics in the Decision-Making Process

YAHIA BENYAHIA and FATIMA ZOHRA HENNANE

*PhD Student, University Ali Lounici-Blida 2, Route d'El Afroun, Blida, Algeria, E-mails: ey.benyahia@univblida2.dz (Y. Benyahia), Hennane_fz@yahoo.fr (F. Z. Hennane)*

## ABSTRACT

The topic of big data is one of the most important research topics in today's world due to its effective role in monitoring future variables and trends in the decision-making of institutions of any kind. In this context, this research paper highlights the importance of adopting big data analysis techniques in enterprises' administrative systems because of their effective role in making accurate and right decisions that will achieve the objectives of the institution efficiently and effectively. This research paper has found that data analysis has become a stand-alone science adding to that there is a great interest from the part of institutions and government sectors to analyze and process big data and extract accurate information to benefit from decision-making.

## 8.1 INTRODUCTION

Today, the world is witnessing major scientific and technological mutations in the field of information systems and technical communication. With the emergence of "big data science" (big data), With its simplified concept, it expresses an enormous amount of complex data that is beyond the capacity of traditional software and computer machines to store, process, and distribute. It is characterized by its large levels of production and circulation in a short and fast time.

In addition to that, its sources are diverse and different from the traditional sources in terms of form and degree of credibility, which made the analysis of these data needs more sophisticated processing systems than the traditional ones. The most important systems are: Hadoop system, SapHana system, CouchDB system, etc.

As rivalry intensifies and customer demand increases, companies and even universities and research centers make efforts to learn the ability and control skills in the analytical methods and tools. Data analytics is a common factor that allows generating insights and patterns and contributes to the development of the decision-making process.

This chapter aims to tackle the following sections: First, the concept of big data, then the characteristics of big data, the next types of big data, later on, big data analysis tools, and finally, the role of big data in the decision-making process.

## 8.2   THE CONCEPT OF BIG DATA

In fact, no accurate definition of big data can be given, as it is a complex and multifaceted term. Wang, Xin, and Aksu have all pointed out that "the concept of big data" first appeared on (Lani) in 2001, describing it as "Data that cannot be processed by traditional data management tools.

Gallagher, Power, and Dollars analyzed 1,437 articles on big data and came up with a common definition that big data is: "The origin of information, which is characterized by large size, speed, and diversity and its analysis requires specific technical and analytical methods to convert them to value; large data comes from the process of accumulation of past and current information about the activity of individuals in various areas of life in order to predict behavior logically or take into account future needs" [1].

In 2014, the United Nations Economic and Social Council introduced the definition of big data as: "Data is large, quantitative, very fast, and very diverse to require cost-effective and framed forms for a deeper understanding and better use in decision-making" [2].

The official definition of big data, according to the McKinsey Institute, is that "a huge collection of data has reached a size that is beyond the capacity of traditional database tools to capture, store, manage, and analyze" [3].

Eric Schmidt, the former chief executive officer CEO of Google, preferred to give a simple and concise definition of big data away from complex scientific and academic concepts. "From the dawn of history to 2003, humanity produced 5 billion gigabytes of data distributed in the form of drawings, documents, and books."

In 2011, 5 billion gigabytes of data were created in just two days, and in 2013, 5 billion gigabytes of data were created every 10 minutes. There is no doubt that the Internet has contributed a lot to achieve this amount of results; the amount of data traded through it is very large, which contributed to the formation of a giant network among billions of people [4].

## 8.3   BIG DATA CHARACTERISTICS

The big data originally is normal data, but it has a set of characteristics that distinguish it from other traditional data, and the most common of these characteristics are known as (3V) represented by (Volume), (Variety), (Velocity), and with further studies, these characteristics were expanded to (7V) and are as follows [5]:

1. **Volume:** The volume of data extracted from a certain source, which determines the value and size of the data to be classified as big data, and by 2020, cyberspace will contain nearly 40,000 megabits of data ready for analysis and debriefing.
2. **Variety:** Data extracted, which helps users, whether they are researchers or analysts, to choose the appropriate data for their field of research, and includes structured and unstructured data such as pictures, audio, and video clips, SMS, call logs, and map data, and requires time and effort to be configured in a suitable format for processing and analysis.
3. **Velocity:** It is the speed of production and extraction of data to cover demand, where speed is a critical element in deciding upon such data, which is the time from the moment this data arrives at the moment the decision is made based on it.
4. **Reliability (Veracity):** It means the reliability of the source of data, and the accuracy, validity, and modernity of those data where we as the Executive Director of all three managers do not trust the data that is subject to the decision. There are also studies that estimate the impact of poor data on the US economy is estimated at 3.0 trillion dollars annually.
5. **Value:** To take advantage of the big data, we need specialists who have the necessary expertise and skills to deal with this data and

analyze it appropriately, in which case the information is considered valuable.

6.  **Variable Value (Variability):** Meaning that the same information or the same data can be customized, depending on the context in which they are presented; their true value can be determined and analyzed appropriately.

7.  **Visualization:** When using big data, it must be analyzed and shown with different shapes to suit the nature of their use, and it takes multiple shapes such as statistics, figures, geometric shapes, and others.

## 8.4   BIG DATA TYPES AND TOOLS

Big data is divided into structured data (or shaped) data, but it presents a small part (less than 10%) and unstructured (or unshaped) data and represents the bulk of the data:

*   **Structured Data:** Is classified, arranged, and stored data in databases, where we can search and extract information from them. For example, ORACL, MySQL.
*   **Unstructured Data:** Is all that can be easily classified, such as pictures and charts, clips of sounds, songs, and videos, click on the sites, web pages, PDF files, e-mails, Twitter Tweets, Facebook posts, chat messages, XML documents, etc.

Although these types of files have a special internal structure, they are considered disorganized because their data is not consistent with standardized columns suitable for a database. Between the two previous types, there are data called:

*   **Semi-Structured Data:** A mixture of the two, but lacking an organized structure, such as word processing programs [6].

The analysis of traditional data is completely different from the analysis of big data. This difference is due to the nature of big data, which is characterized by complexity and high-speed inflation and accumulation in record time, and therefore the analysis process requires more sophisticated processing systems than traditional ones and relying on modern technology among the most important tools used to analyze big data are:

1.  **Hadoop System:** An open-source software, which is characterized by being a framework that allows the processing of large data across a range of computers rely on the work of cloud storage rather than relying on regular databases and simple programming designed primarily for the processing of large data effective processing [7].

    The Hadoop system includes two major systems, namely the Hadoop distributed file system (HDFS) storage system, which is specially designed to handle very large volumes of data, and the MapReduce processing system, which is related to programming languages dedicated to data analysis. C ++, Java, Python, Ruby, R, and other open-source programming languages that are included in the Hadoop package [8].

2.  **SAP HANA System:** This system helps its users to perform simultaneous analysis of big data coming to the system platform, and this enables the commercial and non-commercial organizations that acquire it to perform business processes faster through the availability of processing data to help in decision-making and carrying out the task of planning and implementation with high efficiency [9].

3.  **CouchDB System:** It is an open-source database management system that can be accessed using the JavaScript Object Notation (JSON). The term CouchDB is an abbreviation for Cluster of Unreliable Commodity Hardware; this means that the system is scalable and interconnected with many devices allowing more accuracy and reliability in the data processing. It is worth mentioning that the CouchDB system was initially developed using the C programming language before the process of creating this system with the year 2008 turned to program language Erlang [10].

In general, tools dealing with big data can be divided into three main parts [11]:

- Data mining (DM) tools;
- Data analysis tools; and
- Results display tools (dashboard/visualizations).

Figure 8.1 shows the main differences in the data management and analysis phase before and after the appearance of big data.

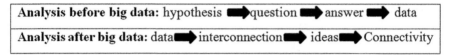

| **Analysis before big data:** hypothesis ➡ question ➡ answer ➡ data |
| **Analysis after big data:** data ➡ interconnection ➡ ideas ➡ Connectivity |

**FIGURE 8.1**    Data management and analysis before and after big data.

Figure 8.2 illustrates the ecosystem of big data in general, which starts from the stage of extracting data from multiple sources through the analysis phase and then the stage of deriving ideas and presentation.

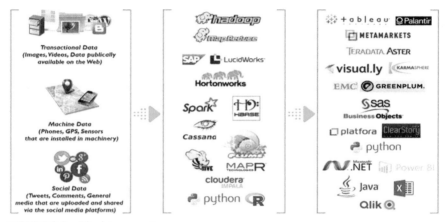

**FIGURE 8.2**    Big data ecosystem.

## 8.5    THE ROLE OF BIG DATA IN THE DECISION-MAKING PROCESS

Big data offers a competitive advantage for institutions that have been able to devise practical solutions to break down their complexity, analyze their content in order to achieve added value and rewarding returns; a good data analysis process leads to a sound, informed, clear and fast decision by decision-makers.

It also leads to faster identification of the appropriate strategy, so data are used to make decisions and monitor progress towards the institution's goals. The value of data in terms of decision-making is crucial and has a significant impact on the survival and development, or the development of institutions or not, and monitor the variables and future trends in decision-making [12].

Big data is important because it enables institutions to collect, store, manage, and handle large amounts of data; thus, obtaining the desired results, large diversity is required, and the size and speed of obtaining such data are very important in light of the development of scientific research methods.

Big data technology also has the ability to analyze sensor data, web sites, and social and behavioral social networking data.

The analysis of this data allows correlations between the independent data set to detect several aspects, such as forecasting business trends for companies, linking legal citations, combating crime, identifying traffic flow conditions, etc. These predictions also provide groundbreaking instruments for decision-makers to better understand clients and markets alike [13].

This new field of data science seeks to extract applied knowledge from data, especially big data that can be analyzed to reveal patterns, trends, correlations, and get insights from them, and ultimately to the scientific implementation of what has been learned. In addition, this field overlaps with all human activities, economics, finance, and business without exception.

Big data contributes to the decision-making process as a result of its special advantages for the business sector, including the following subsection.

### 8.5.1   DEFINING THE FEATURES OF CONSUMERS

The rapid growth in the application of data science in business is not surprising due to the strength of economic considerations in this science. In a competitive market, all buyers pay the same price, and the seller's revenue is equal to that price multiplied by the quantity sold.

However, there are many buyers who are willing to pay more than the equilibrium price, and these buyers maintain a consumer surplus that can be extracted using big data to define the features of consumers.

In addition, charging consumers different prices based on their analyzed features allows companies to obtain the highest price that the consumer is willing to pay for a particular product. Determining optimal price discrimination or market segmentation using big data is highly profitable. This practice has been the norm in some industries, such as the aviation industry, but it currently extends across a wide range of products.

The gains from price targeting also enabled companies to offer discounts to consumers who could not afford the equilibrium price, thereby increasing revenue and expanding their consumer base.

Defining consumers with big data is an important reason for the high ratings of companies such as Facebook, Google, and Amazon, which offer products and services that rely primarily on customer data.

## 8.5.2   FORECASTING AND RISK ANALYSIS

Big data science has been successful in systemic financial risk analysis. The world has become more interconnected than ever, and measuring these links promises a new vision in economic decision-making so that systemic risk is seen through the lens of networks as a powerful approach.

Data specialists are now using abundant data to build images of interactions among banks, insurance companies, brokers, and others. It is clear, for example, that the knowledge of the most interconnected banks will be useful, and the same applies to the information on the most influential banks in the market, which is measured using an intrinsic value-based approach.

Once these networks are built, data specialists can measure the degree of risk in the financial system, as well as the contribution of financial institutions to overall risk, providing regulators with a new way to analyze and ultimately manage systemic risk [14].

## 8.5.3   ARTIFICIAL INTELLIGENCE

Artificial Intelligence is a modern computer science that searches for sophisticated methods of doing works and conclusions similar to those attributed to human intelligence. It is a science that first investigates the definition of human intelligence and determines its dimensions, and then simulates some of its properties by translating mental processes into the equivalent. From calculations, increase the ability of the computer to solve complex problems and make decisions in a logical and orderly manner.

Artificial intelligence is one of the most successful fields at the present time, as it has proven its efficiency in multiple fields and has been applied in many business applications in companies and economic institutions [15]. For example, the latter can make predictions and determine the relationships between economic variables in better and more accurate ways of statistical standard.

It can also be argued that the main reasons for the tremendous success of artificial intelligence are naturally due to two main reasons: the availability of large amounts of data to learn machines, the steady growth in the power of computing, and the development of special-purpose computer chips [16].

The most important applications of artificial intelligence are simulation models, learning algorithms, and artificial neural networks. They can be applied in the business sector, to predict financial market developments, the actions of economic variables influencing the institution's economic climate, and more [17].

## 8.6 CONCLUSION

Big data represents an important stage in the development of ICT systems and has a very promising future for all sectors and fields. This research paper addresses the importance of big data in the field of decision-making at the enterprise level.

In this context, it was concluded that big data analytics are a set of transformative techniques that can help institutions extract value from their data to enable them to develop their decisions. This is due to the multiple use cases of big data, ranging from risk analysis to customer opinion analysis, process analysis, IT security, artificial intelligence, and others, as well as the importance of determining usability and appropriateness to demonstrate the feasibility of investing in huge data.

On the other hand, the big trend in the use of big data does not eliminate the set of challenges associated with it, mainly the need to adopt advanced tools and technologies that can manage and manipulate the huge amount of data, in addition to the need for experts and qualified workers to manage and use these tools.

Through these findings, we can present a number of suggestions, which are summarized in the following points:

- Encourage interest in the use of big data by decision-makers.
- The need to allocate committees of a common nature between the public and private sectors in order to increase research and development in the field of big data.
- Establishment of a national specialized body for data management and governance.
- Apply the Data Protection and Privacy Act in a way that confirms the keenness of actors and their commitment to addressing issues of privacy and safety of users.

## KEYWORDS

- **artificial intelligence**
- **big data**
- **data analysis**
- **data protection and privacy act**
- **decision-making**
- **systemic financial risk analysis**

## REFERENCES

1. Younis, S., & Ahmad, I., (2018). Awareness of the big data concept among academic librarians-case study of the University of Jordan library. *Intervention from the 24th Annual Conference on Big Data and the Horizons of its Investment: The Road to Knowledge Integration* (p. 6). Muscat, Oman.
2. Al, O. M., (2018). Big data in academic libraries in the sultanate of Oman: Reality and challenges. *Intervention from the 24th Annual Conference on Big Data and Prospects for Investment: The Road to Knowledge Integration* (pp. 5, 6). Muscat, Oman.
3. Mariam, L., (2018). Big data and information industry. *Hikma Journal for Media and Communication Studies, 6*(4), 58, 59.
4. Big Data. http://www.snyar.net/143314-2/ (accessed on 22 October 2020).
5. Abdul, R. M., (2018). The role of big data analysis in rationalizing financial and administrative decision making in Palestinian universities: An empirical study. *Journal of Economic and Financial Studies: Al-Ouadi University, 11*(1), 28.
6. Mariam, L., (2018). Big data and information industry. *Hikma Journal for Media and Communication Studies, 6*(4), 59.
7. Hana, K., & Oussama, D., (2017). *Employing Big Data in Technical Companies and User Privacy (Analytical Study of the Usage Agreements and Privacy Policies of Google and Facebook)* (p. 42). Master Note Specialization in Information and Communication Technologies and Society, Faculty of Humanities and Social Sciences, University of 8 May 1945, Guelma, Algeria.
8. Khaled, K., & Saeed, A., (2017). Big data and its impact on the decision-making process. *Journal of Statistics and Applied Economics, National Higher School of Statistics and Applied Economics, 14*(2), 159. Algeria.
9. Ali, B., (2017). *Transforming Big Data into Added Value, King Fahad National Journal, 23*(2), 95. Saudi Arabia.
10. IBM. *Exploring CouchDB,* https://www.ibm.com/developerworks/library/os-couchdb/index.html (accessed on 22 October 2020).
11. Adnan, M., (2017). *Big Data and its Application Areas, Faculty of Computer and Information Technology* (p. 4). King Abdulaziz University. Available at: https://www.kau.edu.sa/GetFile.aspx?id=285260&fn=Article-of-the-Week-Adnan-Albar-01-November-2017.pdf (accessed on 22 October 2020).

12. Sabrina, M., & Mokadem, S., (2018). *The Role of Big Data in Supporting Sustainable Development in the Arab Countries* (p. 06). Presentation of the 24th annual conference on big data and its investment horizons: The road to knowledge integration organized by the specialized Libraries Association, Arabian Gulf Branch, Muscat, Sultanate of Oman.

13. Asmaa, B., & Shaimaa, B., (2018). *Scientific Research in the Big Data Era: Roles and Opportunities for Information Professionals* (p. 05). Intervention from the 24th annual conference on big data and its investment prospects: The road to cognitive integration, organized by the specialized Libraries Association of the Arabian Gulf, Muscat, Oman.

14. Sanjivranjan, D., (2016). Big power in big data. *Journal of Finance and Development, IMF,* 26, 27.

15. Bouzidi, L., (2016). The role of artificial intelligence in forecasting and quantitative risk analysis in the economic institution. *Journal of Economic Dimensions* (Vol. 6, No. 1, pp. 232, 233). University of Ahmed Bouguerra, Boumerdes, Algeria.

16. Sanjivranjan, D., (2016). Big power in big data. *Journal of Finance and Development, IMF,* 28.

17. Latifa, D., (2017). The role of artificial intelligence models in decision-making, *Journal of Humanities* (No. 1, pp. 132). Ali Kafi University Center, Tindouf, Algeria.

# CHAPTER 9

# The Role of the Information System in Making Strategic Decisions in the Economic Institution: Case Study of Baticic in Ain Defla, Algeria

KHEDIDJA BELHADJI[1] and ABDELLAH KELLECHE[2]

[1]*PhD student, Specialization in Production Management, Hassiba Benbouali University, Chlef, Algeria, E-mail: nafoula80@gmail.com*

[2]*Senior Lecturer, Hassiba Benbouali University, Chlef, Algeria, Pb. 02000, Algeria, E-mail: kabd.dz@gmail.com*

## ABSTRACT

The administrative decisions are submitted to the institution according to their organizational levels. However, the high administration is set at the top of the pyramid, which is then responsible for the strategic decisions of the institution and is known as that the quality of its decisions depends only on the extent of information provided by the decision-maker.

In this context, the information basis through which we are trying to shed light on the utmost importance of analyzing data to make strategic decisions by relying on an information system assisting in making the rational decision.

## 9.1 INTRODUCTION

The success of the organization and the increase in the size of its business and activities requires seeking to maintain its position in the market. The rationality of decisions depends on the quality of the information available and availability when needed. Based on this, we can ask the following question:

*How does the role of the information system revolve around making strategic decisions in Baticic that is located in Ain Defla?*

Though, from this question, we can ask the following sub-questions:

- What is the importance of the information system, and what relies on it in the decision-making process?
- What are the strategic decisions, and what are the steps of strategic decision-making?
- What factors influencing strategic decision-making?
- How are data analyzed, and what are their characteristics and importance in strategic decision-making?
- What are the procedures and requirements adopted for strategic decision making at Baticic Foundation in Ain Dafla?

In order to answer the above questions, hypotheses have been formulated as follows:

➢ *H1: An information system is a system consisting of inputs-processed-outputs.*
➢ *H2: Strategic decisions are concentrated at a high administration level.*
➢ *H3: The lack of clarity of the existence of an information system in the enterprise can be considered as factors affecting it.*
➢ *H4: Data are analyzed based on steps designed to make sound decisions.*

To answer the questions and address the subject, we followed the descriptive and analytical approach that enables us to collect and analyze information to reach the results. Therefore, we can categorize the main objectives of the study into three basic objectives, namely:

- Trying to publicize the information system and its importance in the rapid delivery of information, and the extent of its assistance in making sound decisions.
- Highlighting strategic management, strategic decisions, and strategic decision-making steps.
- Directing attention to the need to adopt an information system to analyze data and provide timely information to make appropriate and quality decisions.

For this, we devoted this paper proceeding from the following points:

- Determine the relationship between information systems with data and information, methods of processing, and characteristics.
- Define the strategic management and strategic decisions and factors affecting them.
- Diagnosing the steps, methods, and approaches to strategic decision-making.
- An applied study on the analysis of the relationship between information systems and strategic decision-making Baticic Foundation in Defla.

## 9.2 INFORMATION SYSTEM AND ITS RELATION TO DATA AND INFORMATION: PROCESSING METHODS AND CHARACTERISTICS

The information has become an important resource for institutions due to what has become the reality of current technological development. The manual recording of information now depends on advanced techniques and methods accompanied by high technology with great capabilities in both production processes and in the collection of data.

### 9.2.1 DATA DEFINITION

These are facts that relate to events, whether inside or outside the institution. It refers to these unregulated raw materials and facts, which have no value in their initial [1]. It refers to these unregulated raw materials and facts, which have no value in their initial [2].

This means that data can be considered raw materials obtained through the survey of the internal and external environment of the institution, which was expressed in numbers and values and was used for the first time, such as the volume of sales of an institution.

Regarding the data types, we can indicate that data is divided into two parts [2]:

1. **Numeric Data:** They take specific values.
2. **Representative Data:** Data that takes many unlimited values.

### 9.2.2   WHAT IS INFORMATION?

There are many definitions of information, the most important of which are:

- Knowledge is the meaning and benefit of the individual who is presented to him in achieving his goals, so it has value [3]. Knowledge means information after processing and organized to turn it into an experience that is the final outcome to use the information [1].
- It is the data processed that have undergone conversion and operation to be in a useful form for use [2].
- It is the data that has been prepared to become in a more useful form for its future and has a perceived value in the current or expected use or in the decisions taken [3].
- If we compare data, information, and knowledge, we find that the relationship between these three definitions is interrelated, combining them with an obligatory relationship. The data that is without value at the beginning of their acquisition become after the processing of valuable information, which in turn becomes knowledge for the user in relation to the extent to which they benefit from them.
- When taking on a decision-making process for the decision-maker, it needs many facts that facilitate this process and must have all the facts without distortion or modification; however, information is the basis of the decision-making process, and the information must be accurate and clear.
- And to get them in time, and it must be clear and appropriate for the purpose for which it was prepared, inclusive covering all aspects of the user's interest, flexible to adapt to all needs, and to be diverse and sourced. Alternatively, detailed and be reliable based on the accuracy of the information system inputs in the organization [3].

### 9.2.3   INFORMATION SYSTEM

It is known that any system consists of inputs, processes, and outputs, so the information system is the system through which data is converted into information through operations.

The inputs to this system include data related to the economic events of the organization and external such as sales prices, or internal such as prices of materials used, and then start operations from the moment the data enter the system to convert them such as multiplying the sales quantity in the unit price to determine the value of sales, which is the output of the system [4]. This process includes the following steps:

- Data acquisition in which data from events that occur in a particular form are recorded as purchase orders.
- Validation of the data is validated as a review of the work of another person.
- Classification means that the data elements in certain sectors as sales figures are classified by type of inventory.
- Sorting and ordering any data elements placed in a specific or predefined order.
- Summarize and combine data elements by reducing data such.
- Calculating for the use of data such as mathematical operations required to access the salaries of employees.
- Storing any data placed in storage places in a particular medium such as documents or magnetic tapes, which can be retrieved when needed.
- Retrieving, which requires searching and obtaining any partial data elements from the medium used for storage.
- Reproducing data from one medium to another or in another location of the same mean, such as re-recording from disk to disk.
- Spreading or communicating by any means to transfer data to another place [5].

Regarding the types of information systems, we can notice [5]:

1. **Data Processing Systems:** It collects records and processes data about the daily events of the organization's activities and provides them as information, which expresses routine reports.
2. **Office Automation System:** Serves individuals who deal with data processing for the purpose of speeding up business.
3. **Knowledge Systems:** Serves personnel responsible for creating and operating information in the organization.
4. **Managing Information Systems:** She provides information to managers at the managerial level, which are considered daily or exceptional reports, she gets her information from data processing systems, and she helps monitor current performance and predict future.
5. **Decision Support Systems:** It combines data with analytical models to support decisions. It has analytical capabilities that allow the user to take advantage of advanced models in information analysis and

rely on interaction with the end-user because they work without programmers and professionals' help and provide support for decisions and problems that cannot be resolved in advance. It responds to changing circumstances as per user requirements and relies on information provided by data processing systems, knowledge systems, management information systems, and information from other systems.

6. **Executive Support Systems:** They are designed for managers at the strategic level to support decisions based on internal information summarized and used from management information systems and decision support systems and external information resulting from events in the environment surrounding the institution. It is based on the provision of information to managers upon request as it depends on a general computer system and communication capabilities that can be applied in different situations.

### 9.2.4   DATABASE

It is a structured repository of data related to the files of the institution and the appointment of a set of records of the organization in full. It includes a set of regulated records that are free of duplication, independent of programs, and accessible to all users of the system [6].

According to this study, it is clear to us that the data is the nucleus on which the information is based, but this data, unless processed and analyzed well, will lead to loss of quality. And this is what helps the information system in the institution, which provides support for decision-making through data analysis with the help of technological mechanisms and advanced scientific software.

## 9.3   WHAT ARE STRATEGIC MANAGEMENT AND STRATEGIC DECISIONS AND FACTORS AFFECTING THEM?

### 9.3.1   STRATEGIC MANAGEMENT

There are many definitions of strategic management; the closest to our study is the definition of a crow who praised that the matter in strategic management is related to making major decisions that have a fundamental impact on the future of the institution, as these types of decisions are usually called strategic decisions [7].

### 9.3.2   STRATEGIC DECISION

They are taken by the senior management of the organization and need to allocate a large number of resources, as it is characterized by the direction and future aspiration, because it significantly affects the long-term success of the institution, and must take into account the factors of the external environment [7].

It should be noticed that there are three levels of strategic decisions [7]:

- The first level includes strategic decisions at the level of the whole organization;
- The second level includes strategic decisions at the business unit level; and
- The third level includes strategic decisions at the functional level.

### 9.3.3   FACTORS AFFECTING DECISION MAKING

Factors that influence decision-making are divided into:

i.    External environment factors are concerned with the external environment of the institution, like economic conditions.

ii.   Internal environment factors are concerned with the internal environment of the institution, like organizational relations in the institution.

iii.  Personal and psychological factors: Psychological factors are related to a person's internal motivations, but personality is linked to the personality of the decision-maker and his abilities and behavior [8].

## 9.4   DIAGNOSE THE STEPS, METHODS, AND APPROACHES OF STRATEGIC DECISION-MAKING

### 9.4.1   STEPS IN THE DECISION-MAKING PROCESS

The decision-making process in the institution is not a random process, as it entails strict results that should not be directed to the issuance of ill-considered decisions. In order to make the right decision, the decision-making process must go through the following stages [2]:

- Identifying the problem that has been detected when there is a discrepancy between the objectives and the actual level of performance.
- Identifying and analyzing the necessary data and information.
- Development of alternatives, any solutions or possible means, and available to deal with the problem and solve it.
- Evaluation of alternatives: to know the strengths and weaknesses of every alternative, his revenues, costs, advantages, and disadvantages.
- The choice of alternatives can be based on three perspectives: experience, experience, research, and analysis, but relying on experience is what is adopted as a basis for some to consider it as their guide to know the future.
- Follow-up and evaluate the implementation of the resolution: Once a decision has been made and an alternative selection is made, those concerned are informed of it. It should monitor and evaluation of its implementation by comparing the actual results with the expected objectives of this decision.

### 9.4.2   DECISION-MAKING METHODS AND APPROACHES

Methods of decision-making: It is done according to traditional methods (non-quantitative) and scientific methods (quantitative):

1. **Traditional Methods (Other Than Quantitative):** By amending them to the previous experience using the method of experience and knowledge, personal evaluation following the method of self-government, and the method of conducting experiments by the decision-maker yourself, and the study of ideas and suggestions and analysis in order to choose an alternative solution.
2. **Scientific Methods (Quantitative):** Scientific methods (quantitative): It relies on mathematical models and electronic computers that analyze data and information in order to reach the appropriate decision, such as operations research [8].

Concerning the decision-making approaches, we can mention:

- **President's Control of Strategic Decision-Making:** Assign the chief executive for the foundation of your decision-making. It means that the CEO of the institution to make decisions himself, i.e., the central decision-making, it must be more powerful and influential and has a vision and insight on what to do and how.

- **The Middle Entrance:** The executive officer of the foundation shall initiate strategic decision making with the assistance of his senior subordinates, in order to use them later in the implementation of these decisions, and depends on achieving cooperation and encouraging innovation and contributing opinions and ideas.

- **Delegation to Others:** That is, the president delegated decision-making to others [7].

The information system provides information that senior management needs in order to use it to support strategic decision-making. Management, therefore, uses database systems to provide data that it can use to make decisions at the strategic, managerial, and operational levels, and organize their retention in the computer's internal memory, and facilitate referrals to them when needed. Decision-support systems are also based on the usage of data and help rationalize management decisions.

Generally, different types of decisions demand a lot of information. Management must provide information according to the nature of the work of each administrative level in order to maximize the benefit of the decision-making process because providing the wrong type of information to a certain administrative level is not useful in decision-making and may lead to failed decisions [3].

After discovering the problem, the institution's decision-maker must undertake several steps, studies, and research to differentiate between alternatives, which necessitates continuous research in the institution's environment to obtain new data that are processed in an efficient information system in order to obtain quality information. It cannot be assured that an organization does not rely on an information system that helps it analyze the data. It may cause it to discontinue.

The need for data that produces information in abundant quantity, as the more information, the more the decision-maker is able to differentiate, as well as the availability of the time needed to enable it to be used in a timely manner.

## 9.5    AN APPLIED STUDY IN THE ANALYSIS OF THE RELATIONSHIP BETWEEN INFORMATION SYSTEMS AND STRATEGIC DECISION-MAKING

### 9.5.1    PRESENTATION OF THE INSTITUTION UNDERSTUDY

To understand the problem well and make the appropriate decision in the selection and trade-off between the alternatives available to him, he must have many data processed with high accuracy and speed. These data include the senior management that is competent to make strategic decisions but beyond them to all organizational levels. As an interdependent entity, all efforts are exerted by all levels and branches in order to preserve its position and push it to survive and grow.

The study was carried out at Baticic, in order to know the extent of the institution's reliance on an efficient information system to help it in making its decisions, especially its strategic decisions, which is the public institution for manufactured buildings iron and copper located in the north-west of the municipality of Ain Defla, a total area of 14 hectares, has been independent since 30 January 2005, with a capital estimated at 105,800,000,000 Algerian dinars.

The foundation aims to raise the capacity of producing goods and services in order to raise the coverage rate for the needs of the basic structures and concluding agreements with other local institutions to gradually replace locally made products with imported products, with a commitment to the deadlines, it also has significant material, human, and technological resources.

Work is done through an efficient information system, keeping pace with the development of information, the data is collected in the institution from the external and internal environment, and recorded automatically on computers and then manually in a note organized in tables designed boxes, each field title of the process to be explained. They are stored in classified, structured files, documents, or discs, and after processing, they form valuable information that is circulated in the institution either in writing, such as newspapers and files, orally through telephone or interviews, or through computerized media.

### 9.5.2    THE ROLE OF THE INFORMATION SYSTEM IN THE DECISION-MAKING PROCESSING OF BATICIC

In the intelligence phase and after surveying the internal and external environment, the information system stores large quantities of information that helps in identifying problems and discovering opportunities.

Through the availability of the necessary information obtained previously, can identify the alternatives available to solve the problem and predict the results of their use, in the form of a simple, solvable model that is provided by the enterprise's decision support systems.

For the implementation of the resolution, many concerned parties are contacted, through programmed or emergency meetings, accompanied by previously, designed schedules, which have information on the problem, its causes and obstacles, and expected solutions to it, taking into account the date of its implementation, provided by the approved information system.

The concerned parties shall engage in dialogue among themselves, taking into account what has been recorded in the ancillary tables and plans. Eventually, the concerned parties agree to resolve the problem and issue a decision to implement it, and report it in order to it is implemented.

### 9.5.3   THE ANALYSIS OF THE PRODUCTION DECISION-MAKING PROCESS AT THE BATICIC

To support our study in practice, we have turned to the following question: How to determine the amount of raw material needed for production, for example, in the institution? This is to find out whether the institution uses its information system to determine this quantity, or whether the quantity needed is estimated randomly; we find the following.

The head of the warehouse inspects the remaining quantity of the raw material in the warehouse, collecting all its data of weight, size, color, shape, and type. In this way, information has been collected from the internal environment of the establishment.

The request is addressed to the head of the production, which in turn sends a purchase order to the purchasing department automatically and in writing, for the purpose of executing the purchase order based on the order from the store head. In purchasing management also, the decision is executed after knowing the amount of liquidity available, deciding the method of payment, etc.

The process remains continuous and connected to each other, and this, of course, based on the receipt of the information on a continuous and permanent basis, and analysis and issues the necessary decisions, this is provided by the information system in the institution. Therefore, we were able to answer the question and concluded that the information system has an important role in the institution. It is the basis on which sound decisions are made, and any defect in the system will result in the deviation of decisions and will be characterized by randomness.

The efficiency of all organizational levels in the institution, comparing its actual performance with the expected, appears ineffectiveness of the inter-connection of these levels, and the power of communication and coordination among them. This is reflected in the rapid flow of information between them, and intensifies all efforts in the collection and processing of data to form information to serve all levels.

Data collection is a foundation stone on which any action we accept and which decision we seek to make. For example, creating a factory requires collecting and processing data, forming information to decide on factory size, location, etc., working on data collection.

It is the basis on which most decisions are made, where data supplied from the bottom of the pyramid, beginning with the data gathered by employees, reaches the generation of observations or trends through the stage at which the processing takes place.

In comparison, in decisions involving execution, the data flows in the reverse direction, from the top of the pyramid to the bottom.

Long-term decisions are provided at the strategic stage, for which a strictly internal and external environment must be studied. Data collection is, therefore, a basic aspect in the decision-making process.

## 9.6   CONCLUSION

This research study dealt with data analysis and strategic decision-making by determining the importance of an information system in the organization and its significant role in guiding and supporting its user by establishing the right decision to be made, through obtaining information and obtained after the processing of data.

Data analysis is a colossal step that cannot be ignored because it clearly affects the strategic decision-making process.

Rather, they should be based on studies, researches, and experiments collected from the organization's periphery. Therefore, each institution should use and use information systems to assist in the collection, processing, and storage of this information.

It also makes it available to its users in a timely manner, which is recommended by all institutions with the need to make the necessary changes and adjustments to the programs and mechanisms of operation of these systems, and this is always in order to keep up with developments, and work to survive and continue, but at the lowest cost.

## KEYWORDS

- **database**
- **decision-making**
- **decision-making methods**
- **executive support systems**
- **information system**
- **strategic decision-making**

## REFERENCES

1. Al-Feki, A. I., (2012). *Computerized Information Systems and Decision Support.* Culture House for Publishing and Distribution, Amman.
2. Fred, C., (2011). *Communication and Decision Making.* Treasures of Scientific Knowledge for Publishing and Distribution, Amman.
3. Al-Salem, M. S., (2005). *Fundamentals of Strategic Management* (1st edn.). Wael Publishing and Distribution House, Amman.
4. Nasser, N. A. L., (2006). *Information Systems, Data Processing, and Ready Software.* University House, Alexandria.
5. Al Bakri, S. M., (2004). *Management Information Systems.* University House, Alexandria.
6. Samir, A. G., (2014). *Information Systems Analysis and Design.* Modern Book House, Zagazig.
7. Nabil, M. M., (2006). *Senior Management Strategies.* Modern University Office, Alexandria.
8. Salim, B. J., (2009). *Effective Management Decision-Making Methods.* Al Raya Publishing & Distribution, Amman.

# CHAPTER 10

# The Role of Big Data Analysis and Strategic Vigilance in Decision-Making

BAKHTA BETTAHAR[1] and ABDELLAH AGGOUN[2]

[1]University of Abdelhamid Ibn Badis, Mostaganem, Algeria,
E-mail: bakhta_48@hotmail.fr

[2]Assistant Professor, Faculty of Economics, Business, and Management Sciences, University of Djilali Bounaama, Khemis Miliana, Algeria, Rue Thniet El Had, Khemis Miliana, Ain Defla, Algeria, E-mail: agg88abd@gmail.com

## ABSTRACT

This study seeks to determine the role of data analysis and strategic vigilance in decision-making. Organizations are working hard to improve their performance and increase efficiency to keep in touch smoothly with the latest progress in the field of modern management, where strategic vigilance is one of its most important methods. It brings the organization closer to its environment, makes it known, and help in analyzing big data. In this regard, vigilance plays an effective role in providing the necessary information in many aspects of what affects the organization's competition. Therefore, it helps managers formulate strategies that help them make different decisions that lead to achievement.

## 10.1 INTRODUCTION

The world has recently suffered from many changes and disorders that have resulted from the opening up of local markets, the increase in the volume of big data, and the difficulty of analyzing them, which made organizations live in a highly uncertain environment. This has affected organizations' ability to make decisions, especially strategic ones, due to the large amount of complex information that characterizes the organization's internal or external environment.

The decision-making process requires comprehensive and diverse information to analyze and translate to take profit from them. This can only be achieved if the organization has an important tool: strategic vigilance, which is characterized by its ability to provide accurate and new information in order to carry out what is happening in the external environment and enable it to detect opportunities and threats.

In addition to good knowledge of the internal environment, strategic vigilance is an important way to support information sources and improve competitiveness; therefore, the following problem may be raised:

*How can big data analysis and strategic vigilance which plays a role in the decision-making process within an organization?*

In addition, we can ask the following sub-questions:

- What are the components of strategic vigilance and big data?
- What are the most important stages of the decision-making process?
- What is the prominent role of data analysis and strategic vigilance in decision-making?

In this context, the following hypotheses have been formulated:

➢ *H1: Big data Analysis is a reason for the successful decision-making process.*
➢ *H2: Strategic vigilance information does not contribute significantly to decision-making.*

We have adopted the descriptive-analytical approach, which will help us describe and analyze all the information and data related to the study.

The rest of this chapter is organized as follows. In order to build a vision that helps us to answer the raised problem, we have divided the research into three axes: the first axis includes the components of strategic vigilance and big data; the second axis, the stages of the decision-making process; and the third one, the prominent role of analysis of big data and the strategic vigilance in decision making.

## 10.2    COMPONENTS OF STRATEGIC VIGILANCE AND BIG DATA

The firm is always looking for tools to analyze big data and discover the environment, which will allow it to collect and support information sources and tools in strategic vigilance.

### 10.2.1    *DEFINITION AND TYPES OF STRATEGIC VIGILANCE*

Also known as "a system that helps in the decision-making by monitoring and analyzing current and future scientific, technical, technological, and economic indicators to avoid threats and take advantage of opportunities" [1].

For Reixin [2], vigilance is about "investigating competitors' behaviors, technological innovations, monitoring business strategies, knowing the new desires of consumers and in general, monitoring the environment. All these activities are part of strategic vigilance, which means organizing, testing, interpreting, and disseminating information to improve important decisions in the organization."

It is defined as: "the continuous collective process carried out by a group of individuals in a voluntary manner, they track and then use predicted information about changes that are likely to occur in the external environment of the organization in order to create business opportunities and reduce risks and uncertainty in general" [3].

Rouach defined it as "a set of coordinated methods that regulate the collection, processing, analysis, dissemination, and useful information for the firm's survival and growth" [4].

Concerning the types we can notice:

1.  **Competitive Vigilance:** It is an activity through which the firm identifies its current and potential competitors and all their policies, interested in the environment in which the competing firm develops. The aim is to enable the firm to achieve a competitive advantage to a position in the market, face severe competition in the activity sector, and collect information from a rigorous and strict competitive environment on competitors' movements and activities.

2.  **Business Vigilance:** Interested in customers' needs and desires in the long-term, it also focuses on techniques and how to achieve customer satisfaction, by reviewing complaints that are new approaches to vigilance, and distributors, and sales department. Another area where

vigilance has begun to focus is on human resources by looking for qualified people to hire them and improve cadres' level in the organization by developing a training program.

3. **Technological Vigilance:** Provides sufficient awareness and a new desire among firms' managers and decision-makers to make investments in a process that keeps pace with developments and trends, and enables the follow-up process of Impulsive, regular, and direct monitoring and anticipation rather than wait and only keen to obtain and find access to sources of knowledge and information production.

   It is the process of researching, processing, and disseminating information on technological developments, or with the characteristic of scientific or technical gains, in production and in the packaging process, and allows the registration of patents, technology developments, and manufacturing processes.

4. **Environmental Vigilance:** Monitoring and vigilance of economic, political, legal, and cultural developments that affect a firm's activity; the environmental vigilance allows monitoring: tax developments, financial law, market conditions, changing attitudes, and consumer behavior.

## 10.2.2  THE PURPOSE OF BIG DATA AND THEIR IMPACT

The goal of all this information is how to extract value that enables us to draw full benefit and make decisions through processing and massive data analysis, which is done in several stages: filtering, understanding, and analyzing of this information. However, before the analyzing process, the information must be stored and then viewed, and this can only be through developed mathematical algorithms designed to analyze and handle these data, such as big data analysis, which will help.

For example, in e-marketing operations, customers are targeted with accurate ads about the same product. To do so, some applications gather information about people searching for a specific product, with the potential comparison in many fields with other products.

Big data is the future generation of computing that creates value by data scanning and analyzing. Over time, user-generated data has grown rapidly for several reasons, including purchases at supermarkets, markets, shipping bills, banks, health, and social networks.

With the development of facial and person recognition technologies, it will be able to give more details and information about anyone. As the number of devices that support the Internet increases, devices that are not used to connecting to the World Wide Web, such as cars, refrigerators, and washing machines, all contribute to the increased volume of data produced [5].

Big data can be used in many areas of daily life [6]:

- Governments can analyze the content of their citizens 'social media websites about a particular decision or system that is in place or about to be legislated and implemented and thus knowing citizens' reactions whether they will accept it or not, so it will help in making the appropriate decision for each case.
- Companies can benefit from analyzing data from social media such as Facebook and Twitter to identify their audience and predict the results of their marketing and sales campaigns.
- Musicians can use weblog files and data to determine the details of listening and predicting the popularity of songs in different regions, helping them prepare upcoming live performances.
- Weather, due to the increasing number of sensors we have today attached to smartphones, we have more data than ever before on weather conditions with high accuracy. Today, big data can capture accurate images of prevailing weather conditions in different places in the world and use the information they have to obtain high-resolution weather forecasts.
- Natural disasters prediction, disaster response units can now use data on natural geology and geographic data to predict potential disasters by analyzing past data and then comparing those data to what is currently happening. Thus, these findings reinforce the predictions of these centers and then take precautionary measures before disasters, and natural crises occur, and develop relief and evacuation strategies before it is too late.
- Facebook targets the categories to which ads are directed based on interests' analysis of its users. It uses big data analytics collected depending on information and data provided by Facebook users, messages, user status, and comments, and analyzes all the information contained in the profile of activities, hobbies, age, location, relationship status, favorite movies, songs which will be an important factor in directing advertising and marketing.

## 10.3    STAGES OF THE DECISION-MAKING PROCESS

The decision-making process is a prominent step that many managers seek to master because of the accompanying improvement and progress of the organization.

We can define it as "the process of selecting the best available alternatives, after studying the expected results of each alternative, and their impact on achieving the desired objectives" [7].

It is also defined as "the process of making a judgment about what an individual should do in a situation when carefully examining the different alternatives that can be followed. Or is the moment of choosing a particular alternative after evaluating different alternatives, according to different expectations of the decision-maker" [8].

It is also defined as "one of the manager's principal responsibilities by describing decision-making as an administrative and organizational activity, and the prominent factor is the people who make the decisions" [9]. Alternatively, "the process of a perceived choice between available alternatives in a particular situation, or the process of trade-offs between alternative solutions to address a problem, and choose the best solution" [1].

Through all the above definitions, we conclude that the decision-making process is the one by which the manager selects the best alternative among the available ones that lead to the best solution. It includes the following stages:

- **Identification of the Problem:** The identification of the problem is the first step in decision-making. It is not reasonable to make a decision without an objective. Defining the problem identifies the diagnosis, namely to identify the nature, the dimensions, and the results that caused its effects and causes. This step is very important because any error in the identification of the problem will result in an error in the rest of the phases, which led some to say: "The problem is clearly identified half of the solution and therefore it is desirable that the problem is quantified so, as to facilitate the process of treatment." Also, any mistake at this important stage could lead to making wrong decisions, and the important questions that arise here are: Why was the decision made? In addition, what is the purpose of the decision?
- **Data Collection Information and Data about the Problem:** It is not possible to find a suitable solution to a particular problem without identifying it through the collection of facts and information by an

effective communication system and that information are analyzed to identify the symptoms and causes of the problem, the level of decision-making depends on the validity of this information so appropriate solutions can be found, the information and facts to be collected vary in quantity, quality, importance, and detail depending on the basis of the problem. The availability of information is the main outcome of the decision-maker in devising appropriate solutions.

- **Identification of Alternatives:** The way in which alternatives are developed differs from each other depending on the decision-making method, which can be collective, individual, or participatory. Whatever the difference, the agreement is the key solution, at least in the basic principles. This stage consists of developing possible solutions to a specific problem. The decision-maker relies on his past experience, makes use of the successful elements of previous solutions to similar problems if the decisions are routine. Problems may be partially solved in this way, which part of the solution derives from experience and the other part from the present, and it is better to combine the two parts to obtain an integrated solution. If it is non-routine, it justifies the prominence of the group's creative thought factor or individual decision-maker, predictions about an alternative will be provided to allow a more practical and accurate comparison when considering the best alternative [10].

- **Weigh the Alternatives and Choose among Them the Optimal One:** The evaluation of alternatives is the identification of advantages and disadvantages. According to specific evaluation criteria such as the feasibility of implementation, the impact of the alternative implementation on the organization, the humanitarian and social impacts and implications for individuals and groups, the appropriate time, the extent of response of subordinates, the time is taken by the alternative, in addition to taking into account influential internal and external conditions. This step requires effective forecasting of the consequences of each alternative, and it is useful in reducing types of alternatives after the deleting and omission of alternatives that meet the minimum standards set. Alternatives must also be presented to understand the options that contain a set of more acceptable outcomes that meet the desired objectives [11].

- **Implementation of the Decision:** It is wrong to think that the task of any decision-maker after the adoption of the decision required everything has ended because the decision is not by its approval but with its implementation, and often not taken by the decision-maker and those who implement the decision are usually at the first level of management, workers, and technicians. Therefore, the implementation of the decision is carried out by people rather than those who prepared it, so it is necessary to cooperate, and here comes the role of the organization function, preparation, and assignment of tasks and responsibilities to implement this decision [12].

- **Evaluating Decision Effectiveness:** This stage involves comparing the normative or pre-determined results as objectives with the results achieved to verify whether the decision taken is the solution for the raised problem. The monitoring function's importance is evident during this stage, where the relevant authority decides whether the decision is successful or should be reconsidered, either by modification or by making a new decision, depending on the difference between what was planned and reached. In order to ensure the effectiveness of the decision, the decision-makers will adapt the expected results of this decision to the surrounding circumstances by eliminating all the obstacles encountered after the disclosure [13].

## 10.4   THE ROLE OF STRATEGIC VIGILANCE IN DECISION-MAKING

Strategic vigilance collects, analyzes, translates information, and disseminates knowledge to help make decisions.

The collection of information is a goal of strategic vigilance, which is a key resource that helps the organization to predict and know the changes that occur in the external environment, which helps it to seize opportunities and avoid threats as follows [14]:

- **Control information:** These are information produced by the organization and directed to internal use, and knowledge of this type is very important as it enables the organization to compare its performance with that of the best organizations. This information is managed through information systems such as human resources information systems, production systems, and quality systems. However, they do not represent significant weight for strategic vigilance and are only substantiated information.
- **Influence Information:** Information produced by the organization and directed to external use, i.e., directed to individuals and groups outside the organization (such as the customer and supplier), and

it is managed by marketing information systems and remains only information supported by strategic vigilance.

- **Early Warning Signs (Weak Signals):** Early warning signs represent information that holds the belief that an event that may be of great benefit to the persons in charge of the firm can begin. The warning sign was predictive whenever it was a low-intensity mark and could be expressed by weak signals, the term used by ANSOF. The three types of information can be illustrated in Figure 10.1 [14].

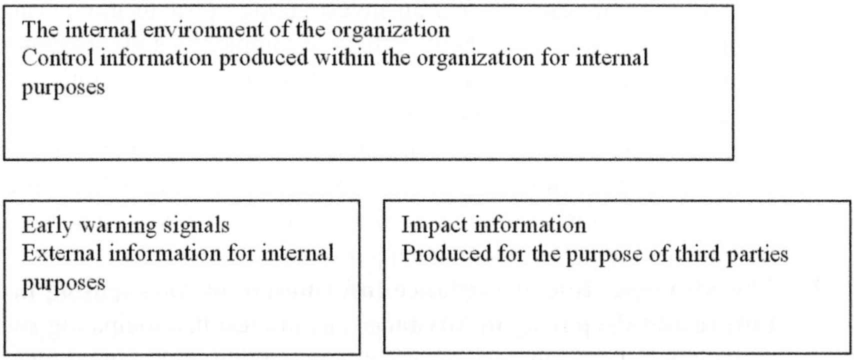

The internal environment of the organization
Control information produced within the organization for internal purposes

| Early warning signals | Impact information |
| External information for internal purposes | Produced for the purpose of third parties |

**FIGURE 10.1**  Types of strategic alert information.
*Source*: Rweibeh, 2003

### 10.4.1  THE STRATEGIC ROLE OF STRATEGIC VIGILANCE INFORMATION IN DECISION-MAKING

Strategic vigilance information is usually linked to the strategic work of the organization, through many strategic areas, the focus of many researchers in the field of strategic management, and although its strategic role varies according to these areas, the importance of this role also differs depending on the effects of its use and the possible consequences. Thus the strategic alertness information plays a prominent strategic role. We can highlight the following points:

1. **The Prominent Role of Vigilance Information in Seizing Opportunities and Avoiding Risks:** Strategic vigilance information is an important factor in the intelligence system of the firm, which gives it greater ability to create and seize opportunities, helps decision-makers to absorb what is happening in the environment, allow for highlighting the offensive side and searching for how to seize the

opportunities, without neglecting the defensive side and the need to the detection of risks and threats. Monitoring the signals of change, recognizing its implications, and extrapolating proactive information reduces risks, exploits opportunities, and makes decisions.

Despite the strategic role of strategic vigilance information, the exploitation of opportunities for some is not just an administrative work done by the decision-maker during the diagnosis of the environment, but philosophy in a life full of risks and spirit of challenge, where the exploiters of opportunities are people distinct from others, with a sense of research and an investigating spirit in monitoring environment events. They believe that the complexity and turbulence that occurs in the environment usually offers many opportunities for those who dare to seize it despite its risks.

In any case, strategic decision-makers should not be preoccupied with thinking, as long as seizing opportunities or avoiding risks require investment in time by reading early relationships and weak signals to anticipate what may happen in the future [15].

2.  **The Strategic Role of Vigilance Information in Anticipating the Future and Preparing in Advance:** The interest in anticipating and anticipating the future was the main concern that took a great place among strategic decision-makers. He monitored the events, followed up on their developments, and monitored the signs that accompany their emergence to anticipate what might happen in the future.

    It is considered as a framework from which predictable trends can be expected, determine its features from the perspective of rational reading of weak signals, inferred on the events of the future events and be prepared before it is surprised by facts, reduces the space and freedom of thought and action away from the natural desire of man to know the unknown to the need head start in the act and make informed and knowledgeable decisions.

    Therefore, strategic decision-makers often resort to modeling to simplify reality and evoke most of the possible perceptions described in some of the dynamic processes, whether observed or simulated, and then interpreted mentally, and develop possible hypotheses in the future. The decision to enter a new market, produce a new product, or withdraw from a specific industrial field, can only be made regarding a set of hypotheses previously formulated [16].

3.  **The Strategic Role of Vigilance Information in Managing Ambiguity and Dealing with Complexity:** The deepening of the vision

and the prolongation of the prediction in the depth of time in order to anticipate the distant future lead to the increase of more ambiguous situations and then increases the risk indicator, this is in addition to the impact of other factors in the composition of events, and the greater the composition of events the situation becomes more complex, and all of this has implications for strategic decision-makers.

Decision-makers fear uncertainty and often make mistakes in dealing with it because of the difficulty of understanding the composition of a large number of environment variables. If future trends are not clear, they describe it as ambiguity. Therefore, the perimeter of the resolution falls between clarity and ambiguity as two extremist points that lead one to the other, where they oscillate between these two points, as the resolution deals with situations and circumstances oscillating between:

- Simple cases, where the composition of the environment is simple; and
- Complex situations, in which events are intertwined, with multiple variables;

Therefore, there must be more accurate and objective methods that help decision-makers to analyze the environment to determine the factors of uncertainty and identify sources of ambiguity [17].

### 10.4.2   THE IMPORTANCE OF BIG DATA ANALYSIS TO RATIONALIZE DECISION MAKING IN FIRMS

The decision-making process is at the heart of the financial and administrative process. The success of firms depends on a large extent of the administrative leadership's ability and efficiency to make appropriate financial and administrative decisions. The decision-making process begins with data collection, processing, storage, and extracting the information on which decisions are made.

From organizations with a large and complex data analysis policy that needs specialized data management and analytics software, which cannot be processed with just one tool or work on traditional data processing applications, The collection of data and information helps to accurately describe the problem and analyze it to achieve accurate results, so it was necessary to adopt a financial and administrative system that includes the analysis of large and very large data.

Many educational firms analyze big data to:

- Improving financial and administrative processes;
- Improving services to the beneficiaries of students, faculty, and stakeholders;
- Development of new educational services;
- Take advantage of appropriate information and provide timely services to beneficiaries;
- Assist in making sound financial and administrative decisions.

The researcher believes that big data is of great importance as it provides a high advantage for educational institutions if they can benefit from it by its processing, storage, and management because they provide a deeper understanding of the beneficiaries and their requirements. This helps to make appropriate and rational decisions within these institutions in a more efficient and effective way, based on information extracted from the databases of beneficiaries and thus reduce costs for students and faculty [18].

## 10.5   CONCLUSION

Under complex environmental conditions, the organization's work can't be without information. It is considered as important as the fuel for the engine. As an open system, the institution can be able to recognize its environment and understand it before making any decision.

Therefore, strategic vigilance is a tool of management based on research, collection, processing, and dissemination of strategic information for use by decision-makers in the institution; it contributes in various types to increase the competitiveness of the institution.

The following conclusions were reached:

- Big data analysis has become a prominent requirement for each institution to achieve its objectives;
- Strategic vigilance is based on a continuous and integrated working methodology for the collection, processing, and timely dissemination of information to decision-makers;
- The process of predicting the effectiveness of the results of each alternative is useful in reducing the offered number of alternatives;

- Strategic vigilance is a system that helps in making decisions through observation and analysis of the scientific, technical, and technological environment. It means monitoring the environment of the institution and monitoring all its changes, and this is in a proactive and voluntary way.

## KEYWORDS

- **big data analysis**
- **data analysis**
- **decision-making**
- **information gathering**
- **strategic vigilance**

## REFERENCES

1. Al-Azzawi, K. M., (2006). *Administrative Decision Making*. Jordan: Treasures of Knowledge for Publishing and Distribution.
2. Reix, R., (2000). *Information Systems and Management of Organizations* (3rd edn.). Paris: Vuibert.
3. Fellag, S. B. B., (2010). The role of strategic vigilance in developing the competitive advantage and reality of the organization in Algeria. *4th International Forum on Competitiveness and Competitive Strategies of Industrial Enterprises Outside the Hydrocarbons Sector in the Arab States* (p. 2). Algeria: University of Chlef.
4. Benhabib, M. D., (2006). *Competitive Intelligence and Decision-Making Tools in Algerian Companies' Case of Service Companies* (p. 56). Les Cahiers du MECAS. N. 2.
5. Lattabi, M., (2018). Big data and information industry. *Hikma Journal for Media and Communication Studies, 6*(4), 61.
6. Al-Ba, A. M., (2015). *Big Data and Its Application Areas*. King Abdulaziz University: Faculty of Accounts and Information Technology.
7. NajmAdallah, J. M. D., (2005). *Principles of Public Administration*. Baghdad: Al-Jazeera Printing Office.
8. Al-Sharqawi, A., (1994). *The Administrative Process and the Function of Managers*. Alexandria: New University House.
9. Shihab, S. M., (2010). *The Ability to Make Decision and its Relationship to the Control Center*. Amman: Safaa Publishing and Distribution House.
10. Khalaf, A., (2008–2009). *The Role of Information Systems in Decision-Making-Case Study of the Naqous Foundation for Pathologists*. Algeria: University of Batna.
11. Qasim, S. H., (2011). *The Impact of Strategic Intelligence on the Decision-Making Process*. Philistine: University of Gaza.

12. Rashad, H. A. S., (2001). *Theory of Administrative Decisions Theoretical and Quantitative Approach*. Amman: Zahran Publishing and Distribution House.
13. RahmanIdris, T. A., (2005). *Business Administration-Theories-Models and Applications*. Alexandria: University House.
14. Rweibeh, K., (2003). A study of the consciousness of Kuwaiti corporate officials towards the use of strategic information. *Arab Administrative Journal, 32*.
15. Yamine, F., (2012 & 2013). *Vigilance and its Importance in Strategic Decision-Making*. Algeria: University of Biskra.
16. Younis, T. S., (2006). *Strategic Thought of Leaders, Lessons Inspired by Arab and International Experiences*. Egypt: Arab Organization for Administrative Development.
17. Cretney, H., (2002). *Management in the Shadow of Ambiguity. Egypt: Future Strategy and Insight.* Abstracts of the Manager and Entrepreneur's Book Series Arab Company for Scientific Media.
18. Rashwan, A. R., (2018). The role of big data analysis in rationalizing financial and administrative decision making in Palestinian universities: An empirical study. *Journal of Economic and Financial Studies, 11*(1), 31.

# CHAPTER 11

# Big Data Analysis and Its Role in Making Strategic Decisions

RAMDHAN SAHNOUN[1] and BOULANOUAR MOKHTARI[2]

[1]PhD Student, Faculty of Economics, Business, and Management Sciences, University of Djilali Bounaama, Khemis Miliana, Algeria, E-mail: laz152.rs@gmail.com

[2]Senior Lecturer, Faculty of Economics, Business, and Management Sciences, University of Djilali Bounaama, Khemis Miliana, Algeria, E-mail: b.mokhtari@univ-dbkm.dz

## ABSTRACT

Algerian economy is facing a serious challenge of coping with the global economy's rapid and dynamic growth while keeping its stability. Due to the consequent rapid flow of big data, digitalization and data analysis became necessary tools used in all economic activities. This chapter attempts to demonstrate the need for big data analysis, inefficient management and examines its role in marking strategic decisions.

The chapter argues that strategic management depends largely on big-data analysis as it might be efficiently and effectively transformed into useful information at the service of the company's decision-makers because it enables them to identify achievable goals at lower costs. The chapter reveals the necessity of transforming big data into measurable information that can be used to assess the company's performance.

Big-data analysis helps the company improve its competitive advantages and gain a larger part of the market while coping with the rapidly changing new data. Since big-data is speedy, huge, and diversified, the company should apply a special system that allows taking risks and investing according to a defined deliberate strategy. Thus, an effective digital system at the service of a company will depend on advanced technology that is updated daily.

## 11.1   INTRODUCTION

Nowadays, man is dealing with a huge amount of different unregulated data. New technologies (smartphones, computers, social media, etc.) provide data that is stored in databases that help organizations and foundations achieve their long-term objectives.

Thus, a company should build a strong qualified infrastructure to benefit from the uses of these new technologies in order to better respond to the speedy flow of big data referred to in the present study as big data. Hence the need to investigate the following research problem:

*What is the impact of big data analytics in making strategic decisions?*
To examine this problem, it is important to answer these two questions:

- How are strategic decisions made? and
- How to draw useful information from big data in order to make strategic decisions?

The study tries to examine the following hypothesis:

➢ H1: Big data refers to the various unregulated data found in social media.
➢ H2: Central administration assigns its sub-departments to access databases and draw useful information.
➢ H3: To make strategic decisions, the company needs to analyze data related to its internal and external environment.

Therefore, this study aims to identify the firm's management's role in analyzing big data to extract useful and suitable information that permits them to make practical decisions at low costs.

The current study sheds light on a very important aspect of strategic management activities: analyzing data in order to draw useful information that allows the company to make crucial decisions and set future attainable goals in a short time and at a low cost.

Recent accelerated events in modern technology have increased the importance of big data analytics for companies of different businesses: commercial, services, health, etc.

Many studies, published mainly in English, have investigated both the theoretical and practical parts of big data analytics. We mention the two following research related to the current study.

Authors in Ref. [1] revealed that data has become bigger than ever before, and data analysis software is more available and practical. The study concluded that the goal of accounting has always been providing internal and external decision-makers with information; nowadays, with big data largely and freely available, accountants just need to collect, scale, and classify this data before transferring it to decision-makers.

The authors of the research about *"The Role of Data Analysis in Rationalizing Financial and Administrative Decisions within Palestinian Universities"* published in 2018 conducted an empirical study about BDA's role in rationalizing administrative and financial decisions in Palestinian universities. The study concluded that storing and processing big data to extract precise information allow making more practical decisions for the Palestinian university [2].

In this study, authors use surveys, articles, and researches related to the topics, mainly those concerned with structuring and organizing big data by experts.

The rest of this chapter is organized as follows. In the first section, we define big data and explore its features. Section 11.2 examines the process of decision-making. Section 11.3 studies how big data analytics impacts the company's strategic decisions.

## 11.2   THE CONCEPT OF BIG DATA

According to Google's CEO, from the dawn of time and up to 2003, humanity produced around 5 Gigabytes of information of different forms: documents, drawings, music, books, etc. In 2011 alone, the same amount of information was produced, yet, in 2013, the same number was produced in 10 minutes, and about 90% of the existing data has been produced in the last two years.

### 11.2.1   DEFINITION OF BIG DATA

"It is the massive speedy various data sets that can be processed and turned to useful information which can be used for a better vision and decision-making" [3].

ISO, The International Organization for Standardization, defines big data as big dataset/s with characteristics (e.g., volume, velocity, variety, variability, veracity, etc.), that for a particular problem at a given point in time, cannot be efficiently processed using current/existing/established/traditional technologies and techniques in order to extract value [4].

It was also defined as "data that exceeds the processing capacity of conventional database systems. The data is big, moves too fast, or doesn't fit the structures of the architecture of your database. To gain value from this data, you must choose an alternative way to process it" [5].

### 11.2.2   BIG DATA CHARACTERISTICS

The most significant characteristics of big data are the 3V: volume, velocity, variety, as called by Dumbill from O'Reilly, and IBM adds another V referring to veracity [6].

It refers to the amount of data extracted from a given topic; the volume and value of this data continue to increase: estimates indicate that about 90% of all ever data created has been produced in the last two years.

1. **Velocity:** It is the speed at which data is created by different categories; everyone is producing a huge amount of information that requires speedy processing to extract relevant information about everyone.
2. **Variety:** Data provided from the use of various smart devices and other sources allows foundations to have access to different types of structured and non-structured data such as sound, drawings, credit cards, videos, SMSs, and telephone calls.
3. **Veracity:** It refers to properties like trustworthiness, relevance, and applicability of events and phenomena impacting the company, and that needs to be analyzed to extract useful true information leading to make as many combinations and associations which permit to make final decisions.

### 11.3   BIG-DATA ANALYSIS

In the past, employees who dealt with accounting, marketing, and other types of documents belonging to the firm's different departments were storing and analyzing data. That was done for commercial and financial transactions and for business information [7]. Nowadays, data is provided by regular people according to their personal analysis. This led big data analysts to adopt a method of analyzing big data for advancing a forecast model (see Figure 11.1).

It should be noticed that big data analytics is done with the help of software utilities like Apache Hadoop, which permits to study of the firm's internal and external environment to facilitate decision making.

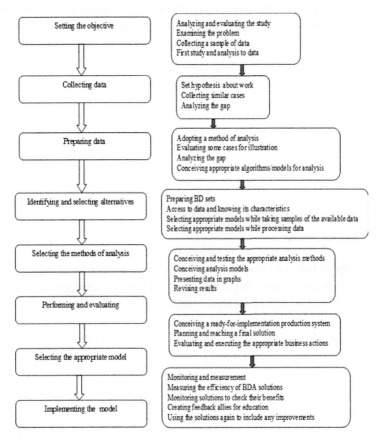

**FIGURE 11.1**   The process of analyzing data.
*Source:* Adapted from Refs. [2] and [8].

## 11.4   DEFINITION OF DECISION-MAKING DATA MINING (DM) PROCESS

The strategic decision strategic decisions play an important role in achieving strategic political goals, in which decision-makers seek to attain different terms: short, average, and long term. These decisions are made according to the goals set by the company, which is made real once achieved [2].

### 11.4.1   DEFINITION OF STRATEGIC DECISION

It is the careful choice between, at least two, two alternatives having the same/similar value. It is an important phase of forming the firm's strategy, which depends on the company's strategic analysis to choose the best. It is also known as the decision made by senior management while responding to the needs of its environment, and which profoundly affects the firm's ability and future.

These definitions conclude that SD aims to pursue the foundation's mission because of the environmental variables. SD is one aspect of the modern organization that seeks to improve its team, risk-taking, and positioning.

### 11.4.2   DECISION-MAKING PROCESS

The necessity of decision marking arises from the existence of many alternatives to carry out the task; opting for the best ones depends on collecting and analyzing financial and non-financial data to make the appropriate decision. This will allow focusing on relevant information to compare the existing alternatives.

The cost of alternative decisions is known as differential costs. Every decision has charges that are borne by the manager and which can be avoided if a given decision is not made. Identifying appropriate costs begins by separating between the variable and fixed costs. The variable costs change according to the level of activity while the fixed do not change.

As the administrative decision is to be taken in the future, its charges are already projected and can be avoided. Based on the aforementioned definition, the process of decision-making involves:

- A problem which needs to be identified and solved;
- Different alternatives;
- Set goal/s;
- Awareness to make a choice.

The process is illustrated in Figure 11.2.

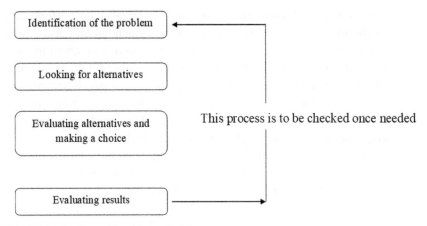

**FIGURE 11.2** Steps of making a decision.
*Source:* Adapted from Ref. [9].

Therefore, the process includes:

- The Identification of the problem: a problem means that the existing situation is not as wanted or desired.
- Looking for alternatives: it is to identify the alternatives and the possible and available solutions for the problem. There have to be two or more alternatives.
- Evaluating the alternatives and making a choice: it depends on the advantages and disadvantages of every alternative, taking into account its expenses and future costs.
- Implementing the solution: in this step, the decision is made and executed.
- Evaluating results: it is to progressively check whether the problem is solved. The previous process needs to be checked.

## 11.5 THE IMPACT OF BDA ON MAKING BIG DATA

With the growing number of users of social media (Facebook, Twitter) and applications such as Gmail and Yahoo, and other web services, large amounts of data have been collected, and many software have been developed to analyze this data to study its impact on performance, competitiveness, and investment in order to make an appropriate practical decision.

1. **Performance Measurement:** Big data can enhance the performance of the management systems as the firm's financial and accounting departments can obtain a format to access information about the size of adjustment with clients, controlling employees' telephone calls, checking e-mails, etc., with the help of data analysis techniques, the information is programmed and used to identify new stimulus measures and internal and external variables, such as measuring employees' morale through e-mails, and telephone calls.

   The culture of making a decision depends on the administration's opinion that is based on sensitive information that needs to be kept confidential in order to foster the evidence-based format.

2. **Competitiveness:** Companies use big data to improve their competitive assets. Huge amounts of data available in social media are analyzed on a daily basis to extract relevant information at the right time [10].

3. **Investment:** Given their characteristics: velocity, variety, and volume, big data are processed using special software that allows the company to take risks and invest at low rates of danger: like investing in the stock exchange where data is huge and non-structured. The investor needs to analyze them in a short time to extract useful information to invest at minimal costs and losses (Figure 11.3).

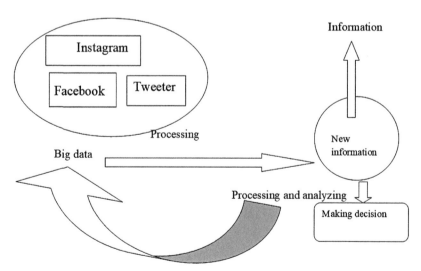

**FIGURE 11.3**  The 3 V's of big data.
*Source:* Authors' elaboration.

## 11.6 CONCLUSION

The study concluded that information systems are rapidly changing; thus, big data are different from traditional data models; mastering their uses and techniques permits the company to improve its efficiency and effectiveness because they are linked to the company's past, present, and future big data is an.

An interdisciplinary field that has multiple sources and they will undoubtedly play the role of the company's neurological system as all other systems will depend on them.

### 11.6.1 TESTING HYPOTHESIS

➢ H1: big data is a set of unregulated data available on social media. Big data concerns people in general, it comes from unspecialized people, and it is analyzed by experts to extract useful information.
➢ H2: the firm's central management assigns its divisions to analyze data and extract useful information. It true, as central management needs an information system to analyze data.
➢ H3: to make suitable decisions, the data related to the firm's internal and external environment should be analyzed. We can confirm this point as it is impossible to make a decision and opt for a suitable alternative without examining the information related to the internal and external environment.

Finally, this study points some results and suggestions:

### 11.6.2 RESULTS

• A repository for archiving big data should be designed and used during planning.
• veracity and velocity should be taken into consideration while analyzing data for an effective and efficient analysis.
• The information system needs to be renewed and adjusted to the characteristics.

### 11.6.3   SUGGESTIONS

- Taking advantage of the new techniques and develop them within the firm.
- Taking into account the consistency of big data analysis.

## KEYWORDS

- **big-data analysis**
- **strategic management**
- **decision-making**
- **information**

## REFERENCES

1. Janvrin, M. W. D., & Watson, (2013). Keeping up with big data. *Journal of Accounting Education, 38*, 3–8.
2. Abderahmane, R., (2018). The role of data analysis in rationalizing financial and administrative decisions within Palestinian Universities. *Journal of Financial and Economic Studies*, 22–44.
3. Goldner, M. T. A. M., (2013). *Libraries' Role Curating and Exposing Big Data, 5*, 429–438.
4. ISO/IEC JTC1, (2015). *Big Data Report*. ISO, Geneva.
5. Dumbill, E., (2014). *Big Data Bootcamp*. CA, US: A Press.
6. Joel, G., (2019). *Data Science from Scratch: First Principles with Python* (p. 1). CA, US: O'Reilly.
7. Yexiong, L. D. A., (2015). Big data analytics and business analytics. *Journal of Management Analytics*.
8. Mesdour, F., (2015). https://w.facebook.com. *Economy and Management*. Facebook [Online]. University of Blida 2, Algeria.
9. Mesdour, F., (2015). *Composer, Steps of Making a Decision*. [Sound Recording]. The University of Blida.
10. Mishra. A. S., (2015). *Information Professional and Big Data, UACSSE, S*(9), 123–129.

# PART III

# Big Data Applications: Business Examples

# CHAPTER 12

# The Farthest Planning of Big Data in the Light of Information Technology: "Smart Cities: A World Not Yet"

NOUREDDINE ZAHOUFI and ABDELKADER DAHMAN

*Assistant Professor, Faculty of Economics, Business, and Management Sciences, University of Djilali Bounaama, Khemis Miliana, Algeria, E-mails: zahoufi.norddine@gmail.com (N. Zahoufi), abd19dah@gmail.com (A. Dahman)*

## ABSTRACT

This study aimed to determine the far-reaching impact of the development of information technology and the enormous digital digitization data in achieving sustainable development through smart cities and highlighting the latter as the main product for achieving sustainable development. Based on the descriptive approach to highlight the concepts related to the study and clarify the relationship between technical information and big data to achieve smart cities, using the analytical method in highlighting the obstacles that lie behind the inability of Arab countries to achieve smart cities.

The study found that access to the building of smart cities requires excellent technological power, high control of the vast data, and the world of electronic digitization. This can be achieved through the convergence of many parties and governmental institutions, social, economic, and security, with the provision of large funding to contribute to the requirements of smart cities' sustainable development.

## 12.1 INTRODUCTION

The 21st century is defined by a tremendous technological revolution, especially in recent years through micro-electronic software, electronic digitization, and the so-called cloud. At the same time, the world knows a terrible increase in the volume of data related to business, population, and the multiplicity of their needs. Economic, social, and service, as well as increased exploitation of resources are characterized by irrational depletion, especially as companies grow around cities.

This has made many international and technological organizations and environmental organizations look into how to control and use this data through information technology, adapting it to achieve a sustainable human life that provides all services, with a healthy life for the individual and ensuring that the right of generations is preserved from Resource depletion.

The aim of the study is to determine the role of information technology and electronic digitization in light of big data in building sustainable cities. Therefore, the question of the problematic study is formulated as follows:

*How do it and electronic digitization lead to sustainable cities?*

The study hypothesizes that it and electronic digitization in the light of big data lead to sustainable cities by building smart cities, which are characterized by a sustainable digital ecology as the result of the combination of massive technological development and the big data of the digitized world, which combines the natural environment with the evolution of the growing digital cloud.

This study is organized as follows. The first section highlights information technology and electronic digitization in the age of big data and its relationship with smart cities. The second section provides a definition of smart cities, their characteristics. A fourth section illustrates the challenges facing Arab countries in the development of smart cities and shows the frameworks for the construction of smart sustainable cities to manage these challenges.

## 12.2   THE CONCEPT OF INFORMATION TECHNOLOGY AND BIG DATA AND THE WAY THEY RELATE TO SMART CITIES

This hub seeks to introduce it and big data and the mechanism of their association with smart cities through the role of information technology as a mechanism for controlling, and using big data and using it in an individual environment.

## 12.2.1   THE DEFINITION OF INFORMATION TECHNOLOGY

Information technology is defined as "the framework for computer science, information systems, communication networks and their applications in various organized humanitarian work."

Contemporary information technology contains five key elements: people, hardware, software, database, and networks, complementing each other, and bonding in a way that makes the system work effectively [1].

These interrelated and interactive elements can be explained as follows:

1.  **People:** Represented by:
    - End users are individuals who use the system or information produced by the system, such as managers, accountants, and customers. On this basis, most of us use the system;
    - Specialists in information systems, which are responsible for the system's operation and sustainability, which technology develops, operates, and manages the information system, including System Analysts, Software Developers, and System Operators from The government's work.
2.  **Hardware:** This includes all the different types of components and physical media used in the processes that pass-through data and information, such as computer systems for various kinds and peripheral accessories in multiple forms.
3.  **Software:** Which includes all and different types of software needed in data processing, including:
    - Operating system systems that help to operate the computer and control its components;
    - Applications, including programming languages, such as V.C+, V. Basic, database software, statistical analysis programs, word processing programs, and electronic table programs.
4.  **Databases:** The data set and documents to be traded within the system.
5.  **Networks:** Which includes communication technologies of various types of networks, such as the Internet, internal intranet networks, external networks/extranet, which have become important in the management of successful e-business, and business processes in all their types.

### 12.2.2    DEFINITION OF BIG DATA

Big data is defined as a collection of a set of large and complex data with unique characteristics (e.g., size, speed, diversity, variance, data health) that cannot be handled efficiently using current and traditional technology to take advantage of. This type of data is accompanied by the provision, processing, storage, analysis, search, sharing, transmission, transmission, and regeneration of this type of data, in addition to maintaining the specificities that accompany [2].

### 12.2.3    THE SUSTAINABLE OBJECTIVES OF BIG DATA IN THE LIGHT OF INFORMATION TECHNOLOGY

The increase in technological development and the so-called electronic cloud and digitization have led international organizations and organizations concerned with information technology, environmental protection organizations, and human beings to research how to exploit the power of big data through information technology in achieving a sustainable environment and cities.

It enables, for the good of the human environment, the use of information technology in the management of big data. This has made it possible to gain access to a variety of environmental benefits by developing what are called cities characterized by high control of big data.

## 12.3    THE ESSENCE OF SMART CITIES

This theme will touch on the concept of smart cities, including the pillars of their characteristics and levels, and clarify their objectives.

### 12.3.1    THE CONCEPT OF SMART CITIES

The term "smart city" is called regional systems with creative levels that combine knowledge-based activities and institutions to develop education and creativity, and digital spaces that develop interaction. Schafer [3] defines the Smart City as a city that uses information, technology, and communications (ITC) to improve the performance of areas as diverse as electricity, water use, parking, and traffic, and waste management.

Cities provide new ways to manage complexity, increase efficiencies, reduce costs, and improve quality of life. Smart cities collect and analyze data and use the information to improve infrastructure, invest wisely, and facilitate the lives of their populations [4].

While the focus group on sustainable smart cities (FG-SSC) defines smart city as an innovative city that uses ICTs and other means to improve the quality of life, operational efficiency, and services in urban areas, competitiveness while ensuring that it meets the needs of present and future, with regard to economic, social, and environmental aspects.

The UK's Ministry of Business, Innovation, and Skills view smart cities as a process rather than a Fixed outcome, where citizen engagement, solid infrastructure, social capital, and digital technologies are combined to make cities more livable and resilient to disasters, Thus, able to respond faster to new challenges [5].

The smart city is based on three main pillars [6], a technical pillar, a social pillar, and an environmental pillar.

1. **Technical Pillar:** It is a digital and virtual city where it is provided with information and communication technologies (ICTs), wireless networks, and sensor networks to form the basic elements of the urban environment and as a system for the operation of the smart community and smart urban management.
2. **Environmental Pillar:** A city that uses new and renewable energy resources.
3. **Social Pillar:** A city focused on the cognitive and creative activities of individuals, knowledge institutions, digital communication infrastructure, and knowledge management.

## 12.3.2   CHARACTERISTICS AND OBJECTIVES OF SMART CITIES

The characteristics of smart cities can be summarized as follows [7]:

- **Smart Environment:** Ensures protection against pollution and management of economic resources.
- **Smart Life:** Includes culture, health, housing, and security.
- **Smart Traffic:** Includes smart public transport and communications infrastructure.
- **Smart Economy:** Promoting innovation, entrepreneurship, and productivity.

- **Smart Governance:** Public services and transparency.

Smart city seeks to achieve a number of objectives, the most important of which are:

- Providing infrastructure for the city to meet the expectations and needs of the population;
- Ensure the quality of life of the population;
- Providing a clean and sustainable environment for the city;
- Providing smart solutions that serve the city and the population in all areas;
- Transforming the city into a "human-friendly city."

## 12.4   REQUIREMENTS FOR THE ESTABLISHMENT OF SMART CITIES AND WAYS OF LIVING IN THEM

To achieve a smart city, the following key requirements must be met:

1. **Smart Buildings:** These are buildings that have integrated systems for managing parts and equipment of the building accurately and quickly and efficiently [8].
2. **Smart Infrastructure:** Smart city application does not depend on the establishment of its vehicles above or underground. Rather, it depends on the identification of additional costs related to subsequent modifications, such as the installation of a layer of sensors and peripherals, which may be directly connected to the Internet and the Internet of things (IoT) [9].
3. **Smart Networks:** The main component of smart cities is the transfer and exchange of data and information between individuals and institutions through applications and networks vary between wired networks, including optical fiber, which is characterized by its enormous data transmission capacity and also a digital subscriber line DSL network that relies on regular telephone lines. There is also wireless Wi-Fi [10].
4. **Being Able to Achieve the Internet of Things (IoT):** The IoT is the use of the Internet to deliver things that generally have the ability to connect to the Internet to send, receive, analyze data and organize the relationship between them in such a way as to perform the required functions and control them through the network [11].

5.  **Global City and Multi-Ethnicity:** Successful smart cities are built on diversity, and talented and creative people prefer to live in cities that are diverse and tolerant and open-minded, stimulating the exchange of ideas and applications of information and encourage the flow of knowledge, allowing the introduction of new ideas that support continuous innovation [12].

6.  **Smart Waste Management:** It works to reduce waste, classify its types from the source, and develop proper treatment methods. These systems can be used to convert waste into a resource and create circular economies. This is through relying on electronic baskets programmed by a type of waste to facilitate their collection and use at the lowest cost [13].

7.  **Smart Energy Management:** Smart grid means gathering information from big power plants to substations on all power grid poles. Once the data is collected and analyzed, power plants can accurately know what is going on in grid networks before it hits the customer, and the generators can be precisely connected to the demand [14].

8.  **Smart Water Management:** It disseminates knowledge and shares it with all stakeholders in water management, as well as making available a variety of ICT resources and infrastructure to build a water management system that exploits aquatic ecosystems in a way that is not prejudicial to social, economic, or environmental sustainability [15].

9.  **Smart Education:** The concept of education refers to the need to learn and apply the principles of ICT in order to make a qualitative leap in the way of learning and teaching, as technical solutions such as virtual learning, digital technologies, and augmented reality change the way individuals learn. Integrated, unrestricted self-education also offers education for all (EFA) opportunities, thanks to data and analysis techniques that contribute to the transformation of digital content-based learning in the classroom into learning through experimentation in the world around the learner [16].

10. **Electronic Health Care:** It consists of the following elements [17]:

- Mobile and virtual health care services.
- Online preventive health services.
- National e-health systems: Is responsible for the issuance of national electronic medical records on a continuous basis and in accordance with local health care systems approved.

- Electronic medical records.
- Health technology innovations.

11. **Dependence on Digital Automation at Work:** Work in smart cities relies on the use of digital automation, where activities that cannot be shortened by rules or algorithms are based on the human element. Although computers, according to strictly defined operations, are focused on the processing of information. This frees workers, leaving the tedious work of computers, towards more imaginative and exciting activities.

12. **Smart City Standards should be Included in Local Authorities' Procurement/Purchasing of Services:** By emphasizing civil and regional aspirations in procurement standards, suppliers can be motivated to invest in smart solutions that contribute to local goals [18].

13. **Cybersecurity, Information Protection, and System Flexibility:** There are several ways to target security attacks and the infiltration of smart city networks. Malware, attack interruption, embezzlement of sent information, impersonation, and internal hackers [19] are the most important. By the full use of available information on threats and computer security incident response resources, situations and expertise at the city level should be established [20].

## 12.5  CHALLENGES OF IMPLEMENTING SMART CITIES IN ARAB COUNTRIES AND STRATEGIES TO OVERCOME THEM

Within this axis, a selection will be attempted to identify the most important obstacles and challenges facing Arab countries in achieving smart cities, while highlighting forward-looking solutions to face these obstacles:

1. **Different National Visions and Their Motives for the Future:** Countries are characterized by a dispersion of visions and ideas toward the future, different systems of the functioning of grass-roots structures overlapping roles, and conflict of interest of tribalism, partisanship, and ethnic affiliation. This challenge can be overcome by the development of a national strategy with clear goals, objectives, and responsibilities that include a general development model.

2. **Diverging, Conflicting, and Resistant Methodologies:** The Arab countries suffer the impact of the dependence of creative behavior on construction and engineering methods. The Arab Maghreb countries

suffer from a struggle in old and modern engineering systems through the French dependency on ancient architecture and Austrian engineering by new engineers. Meanwhile, the government focuses on Chinese engineering to reduce costs and implement the population assembly policy. This challenge can be overridden by developing a governance model for regulation while developing uniform criteria and a structured framework for engineering policies.

3. **Imbalance in Policy Guidance among the Concerned Stakeholders:** This problematic was born out of adherence to office-position- and direct individual management to serve individual interests rather than the public interest; it is the result of selfish thought, which contradicts the foundations of smart cities. This issue can be overcome by setting up a common platform presented in the Smart Cities Office, which allows for the facilitation of communication among the concerned stakeholders to ensure compatibility between smart systems and applications within the Smart City.

4. **Feeble Skills and Behavior towards Modernity:** Arab countries are suffering from the problem of accepting change, especially in the programming and technology sectors. In addition to the challenge of adapting in a different environment to new methods of management and coexistence than they are used to. In particular, workers who are used to not making an attempt to alter the meaning of working and living in another lifestyle. This challenge can be overcome by developing staff programs in Information and Communication Technology (ICT)—in addition to promoting digital knowledge among the population through awareness campaigns, knowledge of the city's smart model, and the way it works and lives in it.

5. **Security Concerns:** Most of the Arab countries have not enjoyed security and stability for a very long time except for a few Gulf countries characterized by semi-stability and security, topped by the United Arab Emirates and Qatar. These two countries know clear progress in heading for smart cities. While the rest of the effects of war and its remnant or the constant fear of its repercussions and the anxiety of seeking security and political stability more than keeping pace with the developed countries and achieving modern models This effect makes it difficult for them to add another security problem of controlling the digital security system, which is still far from it for many years. However, the countries that are heading toward this application, such as the Arab Gulf states, can overcome this problem

by developing policies that guarantee secrecy and privacy through putting digital infrastructures and modern systems responding in ways that counter security concerns to avoid the so-called electronic war.

6.  **High Costs:** The creation and realization of smart cities require a huge financial volume, especially the costs of electronic equipment, digital programs, and cloud software for digital operation. This issue can be easily overcome in many Arab countries by controlling costs, avoiding waste of public money, and making good use of financial resources derived from rentier sectors. In particular, the Gulf and the rest of the petroleum countries, with the condition of exploiting the local resources in construction instead of the random importation of the resources used in the construction of these cities.

    Since the Arab Maghreb countries are characterized by the availability of all the resources for basic base construction, they only have to exploit their own capabilities, and they are left only to import technology and software while trying to keep up with this by encouraging local electronics and cognitive innovation to allow lowering costs.

## 12.6   CONCLUSION

The study found that controlling huge data through information technology, electronic cloud besides its application in the management of resources and the individual's living environment enables the building of sustainable smart cities that are highly resource-intensive, result in social welfare and work to conserve the surroundings and natural environment.

The study also reached the point that the Gulf Arab Countries made a big step toward achieving these sustainable smart cities such as Dubai, UAE city, and Aspetar city in Qatar.

Yet, the rest of the countries are still far from achieving this, especially Arab Maghreb countries, topped by Algeria. This is due to the weakness of modern infrastructures, poor technology related to Information Technology, poor control of technology and electronic digitization, lack of security, in addition to the financial corruption that these countries know during the development projects.

Like smart city-building projects, they need a huge volume of financing, cost control, and strict control of the business. Nevertheless, these elements are absent in the Arab countries, especially the Maghreb, which makes the gap in international progress and welfare grow dramatically in the future when compared to the world of Japan and South Korea.

Through its findings, the following can be proposed for Arab countries, in order to create smart and sustainable cities:

- More use of digital technology in various fields, especially in civil engineering;
- The awareness of the need to apply the characteristics of smart cities and sustainable cities;
- The work on achieving financial governance in the Arab countries to reduce financial corruption in order to channel financial resources to the technology sectors and digital infrastructure;
- Allocate more funding in the digital infrastructure sector;
- Need to educate the community on the rationality in the exploitation of resources to contribute and achieving smart cities;
- Conducting consultative and expert relations with international organizations and countries that have managed to achieve smart cities.

## KEYWORDS

- **big data**
- **digitization**
- **information technology**
- **smart cities**
- **sustainability**

## REFERENCES

1. Bader, I. M. M., (2010). *The Role of Information Technology in Developing Labor Statistics* (p. 5). International statistical course on "developing labor statistics, Ministry of Social Affairs and the Labor Republic of Yemen, Arab Labor Organization Arab Labor Office" Sana'a Republic of Yemen.
2. Al-Bar, A. M., (2017). *Big Data, and Application Fields, Faculty of Computing and Information Technology* (p. 2). King Abdulaziz University, Saudi Arabia.
3. Hanan, E. N., (2018). *Smart Cities: A Study of Concept and Foundations.* Morocco, Law, Morocco. www.maroclaw.com (accessed on 22 October 2020).
4. Hanan, E. N., (2018). *Smart Cities: A Study of Concept and Foundations.* Morocco, Law.

5. Smart Cities, (2015). *Habitat III Issue Papers, 21*, 1. (Abou. Ghazala, Trad). New York.

6. El-Kady, A. N. A. H., & El-Iraqi, M. I., (2016). Characteristics of smart city, and its role in transforming to sustaining the Egyptian city. *International Journal of Architecture, Engineering, and Technology* (p. 2). Iraq no other information.

7. Al-Bar, A. M., & Al-Marji, K. A., (2015). *Internet of Things and Smart Cities* (p. 10). King Abdulaziz University.

8. Al-Aqeel, A. M., (2014). *Journal of Science and Technology Special Issue Smart Cities* (Vol. 28, No. 111, p. 7). King Abdulaziz City for Science and Technology.

9. David, F., (2016). *Infrastructure for New Smart Sustainable Cities* (No. 2, p. 7.). ITU News Magazine, Special Issue Building Sustainable Smart Cities of Tomorrow, ITU.

10. El-Kady, A. N. A. H., & El-Iraqi, M. I., (2016). Characteristics of smart city, and its role in transforming to sustaining the Egyptian city. *International Journal of Architecture, Engineering, and Technology*, 2.

11. Al-Bar, A. M., & Al-Marji, K. A., (2015). *Internet of Things and Smart Cities* (p. 2, 10). King Abdulaziz University.

12. Bahjat, R. S., & Mohsen, J. O., (2016). The role of the information environment in the construction of the smart city. *Energy Journal, 22*, 7. Iraq.

13. Ki-moon, B., (2016). United Nations Economic and Social Council, smart cities, and infrastructure. *Commission on Science and Technology for Development, 19th Session* (p. 8). Geneva.

14. Alice, K., (2014). Smart cities-sustainable cities. *Smart Cities Environment Journal, 8,* 9.

15. Rami, A., (2016). *How Smart is Smart Water Management* (No. 2, Vol. 22). ITU News Magazine, Special Issue Building Sustainable Smart Cities of Tomorrow, ITU.

16. Emmanuel, D., et al., (2017). *National Transformation in the Middle East-Digital Journey* (pp. 29–55). Deloitte & Touche, and Huawei Technologies Middle East.

17. Emmanuelle, D., (2017). *National Transformation in the Middle East Digital Journey* (pp. 23–27). Special Hub sponsored by Future Generation, Deloitte & Touche Middle East.

18. Rick, R., (2016). *Four Ways Political Leaders Can Help Build Smart Sustainable Cities* (No. 2, p. 13). ITU News Magazine, Special Issue Building Sustainable Smart Cities of Tomorrow, ITU.

19. Saleh, M., (2014). Information security in smart cities. *Journal of Science and Technology Special Issue Smart Cities* (Vol. 28, No. 111, p. 35). King Abdul-Aziz City for Science and Technology.

20. Giampiero, N., (2016). *Cybersecurity: A Secure Network for Sustainable Smart Cities* (No. 2, p. 20). ITU News Magazine, Special Issue Building Sustainable Smart Cities of Tomorrow, ITU.

# CHAPTER 13

# Blockchain Technology as a Method Based on Organizing Big Data to Build Smart Cities: The Dubai Experience

SALIHA HAFIFI and FETHIA BENHADJ DJILALI MAGRAOUA

*Senior Lecturer, Faculty of Economics, Business, and Management Sciences, University of Djilali Bounaama, Khemis Miliana, Algeria, E-mails: hafifis18@yahoo.fr (S. Hafifi), magr_fati@yahoo.fr (F. B. D. Magraoua)*

## ABSTRACT

A number of specialists and experts emphasized the importance of keeping abreast of the global development of the use of blockchain technology, which emerged in 2008 and was created as an encrypted series, with various transactions recorded, to be decentralized so that transactions between users are allowed directly without the need for third parties or intermediaries, stressing the need for banks to adopt the Block Qin technology. This Block Qin technology will enhance the performance of banking services and upgrading, and has caused and make a real breakthrough in the world of the digital economy, in addition to sectors such as e-commerce and remittances, smart city infrastructure, real estate, trade, education, health, and other fields.

## 13.1 INTRODUCTION

The world changes rapidly toward urban development, especially in urban areas. That rapid transition, along with the increasing use of information technology between individuals, companies, and governments, will bring socio-economic transformation in parallel change continually into the digital world.

To continue and prosper, cities must deal with and manage population growth, as well as the challenges associated with safety, traffic, pollution, trade, and economic growth, as well as other aspects. Therefore, city officials should ensure an optimal balance between these requirements and meet the citizens' needs and requirements.

Cities also have to manage the expectations of their residents, citizens, businesses, and visitors; they aspire to greater transparency and openness from governments, access to services, and the ability to communicate with the government and in an effort to ensure the building of sustainable, innovative, and resilient cities. An initiative called blockchain technology was launched in 2008 to help transform its green centers into smart cities that live up to the demand.

## 13.2    SMART CITIES AS A SUSTAINABLE URBAN

### *13.2.1    SMARTER CITY: DEFINITION AND CHARACTERISTICS*

IDC research defines a smart city as a limited entity (alive; town; city; province; municipal; urban region) have an authority power at the regional level more than at the state level.

This entity is built on a communications infrastructure and information technology that enables to conduct of a city efficiently and promotion of economic development and sustainability, creativity, and the participation of citizens [1].

The advantages of smart cities are:

- Help to build operational competencies and implement them to provide services to citizens and companies, including electronic services, to obtain approvals and business permits.
- Create an environment that attracts foreign direct investment and maintains economic growth, which contributes to building an urban environment and strong business procedures.
- Support growth, innovation, and accelerating the pace of adoption of technology, such as using citizen's and companies' data to develop innovative new services or applications.
- Ensure a high level of participation of citizens and providing a better quality of life. Smart cities will enable citizens to provide their opinions and observations and communicate directly with authority.

### 13.2.2   THE ROLE OF BIG DATA TO BUILD A SMARTER CITY

The data represents the most important elements that support the success of the city's transformation into a smart city, this transformation is considered successful, and the city should be able to collect data from existing government systems, internet applications, mobile devices, and citizens.

The data collected can be used to make automatic decisions based on proven information, which contributes to improving the lives of citizens.- Smarter cities consist of multiple layers; each containing techniques that assist in data production; classification, analysis, and the ability to respond optimally and effectively, it includes:

- **Connection Layer:** This layer includes all types of communication, such as cellular communication (3G-4G-5G) and Wi-Fi technology, Bluetooth technology. These kinds of connections can be ensured by companies and municipalities, or various government agencies, and make strong communication infrastructure allows cities and access to data and deal with it effectively.
- **Datacenter Layer/Operational:** This layer ensures the data is held in the data repository and deal with it effectively by all departments and applications.
- **Layer Analytics:** Cities benefit through this layer from all the data collected, and to analyze and turn them into valuable insights and activities, cities are increasingly turning to big data analysis that enables them to use predictive analyzes, implementing guidance measures to optimize the allocation of available resources.
- **Layer Applications:** In this layer, cities implement sector-specific applications and applications for different user segments. These applications enable data entry acquisitions and collections across different platforms. This layer provides users with an integrated experience by implementing applications capable of integrating services between different entities.
- **End-Users Layer:** It is the mainstay in any smart city, ranging from individual and public bodies to private companies. This layer is where data is collected through all applications, and it benefits from the products of smart city initiatives.

### 13.2.3    THE ROLE OF SMART CITIES IN THE EXPLOITATIONS OF BIG DATA

Data production is accelerating at an unprecedented rate; the data world is expected to reach 44 zettabytes by the year 2020 [2].

Cities recognize that becoming smart cities requires increasingly big data technology, to be activated across different data entry channels, enabling them to derive good insights and ensure a strategic decision-making process. Also the city's ability to respond to information, data collection, and protection it is essential for her to become an information-based city in order to make the transition to smart cities the following aspects of data need to be considered:

- **Data Source Integration:** Cities will need to ensure that data is obtained from various source, and cities will need to determine how data is classified and archived and to determine the levels of rights of different public and private entities to access data in the early stages, cities will have manage issues related to data interoperability by different administrations and manage their quality and formulas, cities should also link and cooperate between old and new environments and data being produced.
- **Big Data Analytics:** Cities will be able to ensure valuable insights by leveraging analytics, and big data technologies provide predictive analytical solutions to users and businesses and measure performance.
- **Governance Information:** Governments will have to deal with how data is being managed, to balance the data with its confidentiality and access.

### 13.3    BLOCKCHAIN TECHNOLOGY AS A METHOD TO ORGANIZE A BIG DATA

### 13.3.1    DEFINITION OF BLOCKCHAIN TECHNOLOGY

Blockchain emerged in 1968. It was created as an encrypted string, where transactions are recorded I Bitcoin virtual currency conversion, to be decentralized design (see the patterns of the decentralized network in the center of Figure 13.1 dating back to 1967 when Paul Bara published it illustrate the system of distributed centralized and decentralized systems) so transactions are allowed to be directly between users without having to have third parties [3].

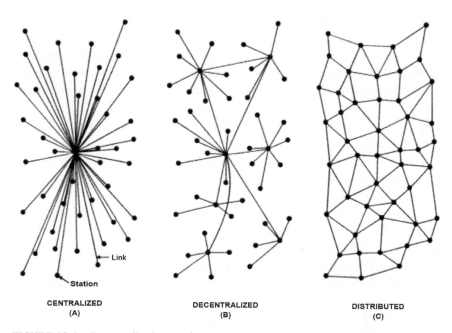

**FIGURE 13.1**    Decentralized network pattern.

*Source:* Ahmed, 2018 [4].

Blockchain technology has created real infidelity in the digital economy, as well as various sectors such as E-commerce, remittance, and smart city infrastructure, real estate sector such as property registration, commercial exchanges documentation of equations, brokerage, education, health, and other fields.

Blockchain is the largest distributor and opens digital record that allows the transfer of the asset of ownership from one party to another at the same time in real-time, without the need for a broker, with a high degree of security for the conversion process in the face of fraud or manipulation attempts. Everyone around the world participates in this register. Blockchain can now be considered the largest database distributed globally among individuals [5].

Blockchain technology is a long series of encrypted data distributed to millions of computers, and people around the world, allowing many parties to enter and verify the information [6]. Each computer or entity in this series has the same information, and if part of it is broken or compromised does not affect the rest of the series [7].

Blockchain technology is based on two basic concepts: (i) a business network and (ii) a ledger thought which members exchange valuable goods through the ledger; each member has its content and agreed with others. Blockchain is an open-source, decentralized database based on mathematical equations and cryptography to record any transaction or information, such as cash transactions, goods transport, general information, or even electoral votes.

Blockchain is a term for the process of producing successive blocks in a virtual configuration process that is modified sequent any accounting record of the year in the financial sector and is currently exploring another user of it in many other sectors such as the logistics sector as taking the delivery of goods and trashing their progress.

Blockchain assets in virtual currency, it was a database to track online mining traffic. To calculate the virtual currency of each user, the mining and bitcoin extraction in this database was documented as a series of documented blocks extracted worldwide. It is, therefore, impossible to falsify a mass or add another untrue to the public register blockchain without being approved by all concerned parties and operating on the internet [8].

It should be noticed that Bitcoin, which is an encrypted digital currency, can be traded in a decentralized manner without relying on a central source such as banks to issue the currency and monitor its prices. Initially, many bitcoins could be purchased in one-dollar until, in October 2017, the value of bitcoin per bitcoin rose to more than $6,000 to set a new record to double its value in 2017 only more than six times the beginning of the year.

However, in reality, it is not just innovation in bitcoin; it's in the technology that Bitcoin originated on, which is the blockchain [9].

According to the latest available reports, there is a shift of 15% of the largest banks in the world in the launch of products based on the technology of SharePoint IBM announced, which has investments in the IoT which works with technology Blockchain about 2000 million dollars, the global market value of blockchain technology is expected to reach around 20 billion dollars in 2024, technology, and financial companies invested 1.4 billion dollars in blockchain technology in 2017.

In addition, about 90% of global banks in Europe and North America are exploring an investing in Blockchain technology, which can reduce bank infrastructure estimated to 30% [3].

In 2013, Vitalique Putin introduced Ethereum, which is based on block-chain technology but not only as a virtual currency but also as a currency that offers smart contracts to eliminate the need for a third party.

The idea of smart contracts in the use of blockchain technology is to document information between parties within the database and is open to viewing. Smart contracts replace the third party where you first document the ownership of the first person of the place of the sale by reviewing the contract book registered in the blockchain and then verifying that the second party owns the required value of the object in the sale and then documents the transfer of ownership from the first party to the second party in DFTT contracts constantly updated with the adoption of blockchain technology [7].

Bitcoin is the world's first virtual currency that does not need a central party to control it, as is the first currency through which smart contract soft-ware can be developed. Blockchain uses to open up prospects for a new type of governance.

### 13.3.2  ELEMENTS OF BLOCKCHAIN SYSTEM AND PRINCIPLES

The blockchain systems consist of four main elements are [5]:

- **Mass:** Represents a chain building unit, a set of processes or tasks to do or to perform within the series, is an example of blocks transferring funds, recording data, or monitoring a case, and usually absorbing.
- **Information:** It is the sub-process that takes place within a single block.
- **Margin:** Is the DNA characteristic of the mass chain, sometimes symbolized by a digital signature. It is a code that is produced through an algorithm with a block program called a Margin mechanism.
- **Fingerprint Time:** It is the time when any operation was performed within the series.

The blockchain system operates according to three basic principles [5]:

1. **The First Principle: Open Log:** All information within the Block-chain is available to all individuals within the chain, are each other's property but retain the inability to know their true identity. This is because the series allows individuals to use nicknames other than their real names.

2. **The Second Principle: Distributed Data-based:** This principle aims to eliminate the idea of centralization. There is no single destination; server, or device controlling the cluster chain; anyone in the world can download, view, and salve the series. This principle is one of the safety elements of a chain.

3. **The Third Principle: Mining:** Mining means the use of computer energies to find the right margin characteristic of this work until it is successful; millions of prospectors around the world perform a series of complex calculations across their devices to obtain the correct margin that links this transaction to the previous transaction within the chain, this is the main function of the mining process it is to ensure that the new transaction took the same period of time as the previous transactions within the chain to ensure that no manipulation or fraud, once the correct margin is obtained the transaction is completed and allowed to enter the chain and is joined to other processes within the block formed at the end of the blockchain.

### 13.3.3　REASONS FOR BLOCKCHAIN DEPENDENCE AND APPLICATIONS

Commercial transactions between individuals are done through intermediaries. The latter receives a percentage of transactions as fees or wages for mediation. Blockchain technology helps in any transaction or transfer of life to another party, storage, and management without the need for an intermediary.

What makes blockchain one of the engines of the intelligent revolution in human life and makes it one of the most important tools for managing the lives of individuals is the availability of two main advantages that can be summarized as follows:

- The main objective of blockchain is to transfer the origin of the object to another party via the internet; what always happens is the transfer of a copy of the life, not the transfer of the original file, which can't happen when trying to transfer the origin of something like money.
- Protect the transaction from manipulation: Blockchain has the important characteristic of making sure not to cheat and fraud during the executions. Thus, the blockchain prevents the manipulation of transactions in such a way as to damage the state's wealth or violate the principle of equal opportunities. This helps to eliminate corruption.

The blockchain system ensures that it is not manipulated, not modified, or delete later; in the case of a problem, the network is able to correct itself to ensure the validity of the transaction protection of its data by a mathematical equation called proof of work [7]; this helps create trust between users in a big way.

Concerning its applications, we can mention [5]:

- **Property Registration:** One of the functions of the blockchain system is the ability of individuals to register their property whatever this property, be it real estate, land, jewelry, precious stones, cars, personnel property, patents, and intellectual property rights is books, songs, notice or other property owned by individuals and wish to advertise it or register it to ensure their rights.
- **Documenting Information:** Any transaction, whether personal between individuals or within a company or a governmental or non-governmental institution, is intended as an open and distributed digital record, allowing everyone to enter all data on it, whether this data is government procedures, or follow-up production lines in a factory, aircraft itinerary or carrier. Oil, as well as the registration of sales and purchase transactions, transfers of ownership, follow-up customer service, and registration of all transactions made between two individuals in any field, allowing the detection of gaps, anti-corruption, and quality control.
- **Brokerage Business:** The blockchain plays the role of an intermediary during the provision of the service banks in the transfer of real estate money, in the registration of property and the place of traffic departments, in the registration of cars and the shop brothers and selling and intermediary companies like umber in providing services.

### 13.3.4   BLOCKCHAIN RELEASES AND FEATURES

Authors in Ref. [10] indicated that there are major versions of blockchain, and each has a workflow is: hyper ledger fabric-R3-Corda-Ethereum.

Blockchain releases vary, and many companies and institutions want their own publications that are not available to the public on the internet, but to specific categories, but there is a great risk as this protection becomes a difficult process because blockchain in bitcoin, for example, contains tens of thousands of people who manipulate it by importing their many hard to manipulate records.

In addition, there are many advantages of blockchain, and the functions that can be performed. It is an administrative and financial system capable of carrying out several real functions while saving as much time, effort, and cost of doing that. The most important features of the blockchain system are the following [6]:

- **Face the Routine:** This system helps government departments to achieve efficiently, all transactions for individuals are visible within the chain if there is a need to make sure some information can be accessed easily, which helps save time and eliminate routine.
- **Anti-Corruption:** The system does not allow Blockchain modification, or cancellation and all transactions that contribute to it are recorded step by step. In the case of manipulation or forgery, the chain does not accept the introduction of the transaction again to helps eliminate corruption.
- **A Fair Distribution of Wealth:** This system contributes to the distribution of wealth among, all individuals around the world and not monopolized by some bodies or organizations, because everyone around the world can participate in the completion, and preservation of transactions and get a percentage of them, in addition, blockchain can free large numbers of people around the world from the bank's power and the cost of using it [11].

## 13.4   EXPERIENCE OF DUBAI

Dubai in the UAE is at the forefront of the transition to a smart city, and the goal of these smart initiatives and strategies is to ensure "happiness" for citizens and provide everything they need.

The mobile phone is the main technical driver of the transformation of Dubai into a smart city in light of the high penetration rate of mobile phones in the Emirate to 200%.

### 13.4.1   DUBAI GOVERNMENT'S STRATEGY IS BASED ON-PILLARS

Infrastructure, transport, telecommunication, financial services, urban planning, and electricity, as well as water initiatives being launched across these sectors [11].

The success of Dubai smart government depends on three key factors: communication, and data communication technologies. The first phase of transformation was completed in 2017. The second phase and the main driver of this transformation will continue by 2020, when Dubai hosts the world expo.

The Dubai Roads and Transport Authority announced in early 2015 that nearly 100% of its services have become smart, and with authority providing about 173 services via the internet and mobile platforms for public transport and business users, it has also established a modern data center that will enable it to collect data. From various services and the establishment of an integrated control center [11].

In 2016, The Dubai Electricity and Water Authority launched its first initiative by deploying 200,000 smart meters, allowing citizens to monitor their electricity and water consumption, helping them reduce their consumption.

The plan is scheduled to include the deployment of more than 1 million sensors by 2020. Dubai Electricity and Water Authority are deploying a smart solar energy network and the construction of fuel supply stations for hybrid vehicles to confirm its commitment to improving sustainability [11].

Dubai police also deployed about 650 television cameras to monitor commercial sites. Data helps police from these cameras and reduces the time taken to resolve the issues [11]. Dubai has established two major projects [12].

Dubai district and Dubai silicon park, to reinforce the Emirate's commitment to becoming a smart city all initiatives designed to make Dubai a smart city rely heavily on blockchain technology, which the government aims to deploy to deliver an unparalleled experience for citizens and businesses [13]. The government of Dubai has pledged to upload all its documents on blockchain platform by 2020.

Dubai took the lead position of global efforts to be the capital of blockchain during 2018, and within a short time of launching the strategy of the blockchain was developed more than 20 cases of the use and employment of blockchain in government services [14].

## 13.5   CONCLUSION

Blockchain technology is a new paradigm shift through which dilemmas can be solved that force organizations to adopt the central system, the centrality of the systems and the need for them is based primarily on the need for a party that can be trusted that can preserve the rights of the clients and can ensure the effectiveness of A millimeter.

But with blockchain, the need for a trusted party is no longer important as instead of trusting institutions, governments, or countries on which to trust the technology and mathematical equations on which blockchain is built. These equations are impenetrable, modifiable, or altered and continue as long as the network continues to operate, and at the same time, these equations are open to all to be monitored, which reinforces the principle of control and transparency.

Technology is still in its initial stages, making it vulnerable to criticism at times and sometimes to fear for its application. However, the solutions it offers can break concepts that have existed for decades as the concept of centralized money and the need for a central bank to regulate the operation of currencies or the concept of the need for the authority of the central state, which presents itself as a third party to resolve disputes between citizens and the possibility of citizens trusting them.

## KEYWORDS

- **blockchain**
- **data**
- **Qin block**
- **smart city**
- **technology**

## REFERENCES

1. MICACOMAR, (2015). *Build Smart Cities Based on Smart Data* (p. 2). Technical Docs ADS.
2. http://www.emc.com/leadership/digital-universe/ (accessed on 22 October 2020).
3. Al Watan Economic Newspaper, (2017). *The Number 3478*. p. 4.
4. Ahmed, S., (2018). *Guide of Block Chain for Start-Ups* (p. 4). BIRM.
5. Ihab, K., (2018). Blockchain, the next technological revolution in the world of finance and management. *Future for Advanced Research and Studies, 3,* 1. Abu Dhabi-UAE.
6. Mega, K., (2018). *Blockchain: The Heart of the Global Financial System,* 28. On the site: http://www.entrepreneuralarabiya.com/ (accessed on 22 October 2020).
7. El-Nemr, M., (2017). *Blockchain: Towards New Horizons of Governance* (p. 4). Egyptian Institute for Studies.

8. https://www.oreilly.com/ideas/understanding-the-blockchain (accessed on 22 October 2020).

9. El-Nemr, M., (2017). *"Blockchain": Towards New Horizons of Governance* (p. 15). Egyptian Institute for Studies.

10. https://www.oreilly.com/ideas/understanding-the-blockchain (accessed on 22 October 2020).

11. Mica, K., (2015). *Building Smart Cities Based on Smart Data, IDC Technical Documents*, 16.

12. *Does Blockchain Technology Revolutionize the Economy of the Middle East?* On the site: https://alghad.com/ (accessed on 22 October 2020).

13. The United Arab Emirates, (2017). *The Future of Money Planning for the Future-E-Money Network* (p. 12). World Government Summit in collaboration with EY, your government supported by blockchain technology.

14. *The Blockchain Technology Forum Opens in Dubai begins in Dubai on the Site.* https://www.albayan.ae/?webSyncID (accessed on 22 October 2020).

# CHAPTER 14

# The Uses of Big Data in the Health Sector

FATIMA MANA,[1] REDOUANE ENSAAD,[2] and DJAZIA HASSINI[3]

[1]Senior Lecturer, Department of Management Sciences, University of Hassiba Ben Bouali, Chlef, Algeria, E-mail: f.mana@univ-chlef.dz

[2]Department of Commercial Sciences, University of Hassiba Ben Bouali, Chlef, Algeria, Pb. 02000, Algeria

[3]Department of Economic Sciences, University of Hassiba Ben Bouali, Chlef, Algeria

## ABSTRACT

We live in the modern era, among large collections of big data swimming around us. In order to develop and facilitate our lives, we must pay attention to the analysis of big data because it is an important resource that helps the various bodies interested in finding solutions to problems and crises that may afflict institutions. This study is therefore in the context of academic research that looks at the development of the health sector through the use of big medical data analysis and adds modern scientific value to previous research that has focused on the study of big data.

The theme of the paper is to clarify the importance of big data processing and its uses in the health sector, focusing on the definition of big data and on the identification of its characteristics, and on how to use it in the health services sector, citing the French agency (ATIH). It also aims to give recommendations to officials on Algeria's health sector, which are intended to lead them to pay attention to the treatment of big data because of their benefits, such as European and American health institutions.

## 14.1 INTRODUCTION

The transition from an industrial society to an information and knowledge society affects the social, economic, and cultural aspects of individual life, as there are currently few aspects of life that have not been affected by an information system. In recent years, information technology has become a reality and has an active role in changing the economy and business, as businesses are being operated in a global environment and cannot be accomplished without information based on automated media.

In order to become acquainted with the subject of research, we divided the study into the following topics:

- The concept of big data;
- The characteristics of big data;
- The employment of big data in the health sector;
- The way in which big data is processed at ATIH France;
- Conclusion, limitation, and future research.

## 14.2   THE CONCEPT OF BIG DATA

Big data in its origin is ordinary data. Still, it is characterized by the large size, i.e., quantity, and characterized by the speed in terms of its composition and generation, and the multiplicity of its forms. All these qualities, and others have created difficulty in processing this data and how to store it and convert it from raw data to information of cognitive value is the essence of making use of this data. Before you start defining big data you should know the meaning of data.

### 14.2.1   BIG DATA DEFINITION

The term "big data" appeared in 2000, and the term refers to the raw material of information prior to sorting, arrangement, and processing and cannot be used in its initial form before processing. Information is data that have been processed, analyzed, and interpreted and can be used to develop different relationships between phenomena and decision-making, and raw data can be divided into three types [1]:

- **Structured Data:** Data organized in the form of tables or databases for processing.
- **Unstructured Data:** The largest proportion of data is the data that people generate daily from text, images, videos, e-mails, etc.

- **Semi-Structured Data:** It is a type of structured data, but the data are not designed in tables or databases.

Therefore, big data is seen as a hard-to-define term, because the concept of 'mega' refers to the large volume of diverse and different data from one institution to another, which does not mean that it is a technological combination, on the contrary, defined as a type of technology and techniques [2].

Although the term is not widely used in other definitions of big data, it refers to new use in which many fields converge: statistics, technology, databases, and professions (marketing, human resources management, finance, etc.).

This new use was made possible by the power of technology that made things possible after it was not far away, just a theory; these things are related to two factors: the size and complexity of the data [2].

In another definition, which describes big data as data of different sizes, of varying complexity, generated at different speeds, and multi-diversity, cannot be processed using traditional techniques, originating from sensor networks, nuclear plants and X-ray scanners, Scanners [3], aerial monitors, and social media such as phones, Facebook, Twitter, and others.

As for the economic institution, the major data are inputs for an information system that processes them like any other system to produce them as outputs in the form of usable knowledge information, so these inputs have sources from which they collect, two types [4]:

1. **Internal Sources of the Organization:** In the form of e-mails, main server records, blogs, and documents, business events, and any other regular, irregular, or semi-regular or semi-regular data.
2. **External Sources of the Organization:** As the institution has a lot of external data, some of which comes from the subscriptions paid in the information banks where this data is purchased, and the rest is available selectively to the organization and sourced by specific commercial customers, social networking sites and protection associations. Consumers and other parties with direct or indirect interests with the enterprise.

## 14.3 THE CHARACTERISTICS OF BIG DATA

The researchers disagreed in identifying the characteristics of big data as they disagreed from other traditional data. The term big data indicates that this type of data has features distinguishing it from the rest of the data, which is the same feature for which the data was formed.

Some of them are limited to three, some of them are limited to four, and some of them are more than that. But the most basic features of big data are what have been described in previous definitions through which we extract these features.

### 14.3.1   VOLUME

We cannot imagine the magnitude of the data no matter how much we use from a unit of measurement unless we show how much these units are equivalent to the amount of data flowing, and as with the weights used in measuring physical and liquid goods, we use kilograms in weight, meter in the calculation of length, etc.

In calculating fluid volume and so on, the data is measured in a unit called Bytes, and every 1 Kbytes equals 1024 bytes, and Table 14.1 shows the data measurement units.

**TABLE 14.1**   Units of Data Measurement

| Unit | Measure |
|------|---------|
| Bit | 1 or 0 |
| Byte | 8 bits |
| Kilobyte | 1.024 bytes |
| Megabyte | 1.024 Kilobyte |
| Gigabyte | 1.024 Megabyte |
| Terabyte | 1.024 Gigabyte |
| Petabyte | 1.024 terabyte |
| Exabyte | 1.024 Petabyte |

*Source:* Adapted from: http://www.wu.ece.ufl.edu/links/dataRate/DataMeasurementChart.html.

The researchers are similar to the increase in data in the tsunami data, where statistics indicate that as of 2003, 5 Exabyte was created from digital data, while in 2011, the same amount was recorded in two days, and the same amount was expected to be recorded every 10 minutes of 2013 [4].

Table 14.2 refers to the explosion in the flow of big data by some means of mass communication [5].

**TABLE 14.2**   Big Data Flow in 2015

|  | E-Mails | Mobile Phones | Search on Google | Telescope | Facebook | Twitter |
|---|---|---|---|---|---|---|
| Data size | 145 billion messages | 450 million messages sent | 4.5 billion searches | 30,000 megabyte of data collected | 552 million users | 400 million users |

*Source:* Adapted and modified from Eric [5].

In 2012, the French newspaper "Le Monde" published an article on big data: every second, the amount of data generated on the network from mobile phones, mobile devices, and electronic panels in the form of images, videos, texts, messages, and numbers, originating from individuals and institutions. Every day, it sends 118 billion messages; 2.45 billion texts are sent to Facebook, the U.S. telecommunications customer sends 240,000 billion octaves of various data online every day. The paper adds that humans are not the only ones who send information; the machines also participate in the flow of data through devices connected to each other, millions of information flow from the SIM card implanted in computers everywhere (car computers, petrol stations, satellites Industrial) [6].

## 14.3.2   VELOCITY

Data speed has two aspects: speed related to the volume of data flow, which means the speed of production and extraction of data to cover demand, and here the need for a system that ensures superior speed in the analysis of big data in real-time or speed of near-instantaneous time. This need has led to the creation of technologies and solutions such as Apache, SAP HANA, Hadoop, and many other programs.

The data transfer rate is only a data transfer time unit, the capacity that a system allows to transmit data over a connection, and this quantity is measured in a byte unit for each time period, usually per month.

Speed is also defined as the way in which data runs in any direction (transmission, reception), which is limited, and the larger the better, and plays a key role in increasing or reducing the physical components of the conveyor wire [7].

### 14.3.3  VARIETY

It means the diversity of data extracted, which helps users, whether researchers or analysts, choose the right data for their research field and includes structured data in unstructured databases and data that comes from its unsystematic nature, such as photos, clips, audio recordings, and tapes Video, SMS, call logs, GPS data and many more, and require time and effort to prepare them in a suitable format for processing and analysis.

In addition to what is known as the traditional (3V) properties of mega data, other data equally important in the description of big data have been added.

### 14.3.4  VALIDITY

Unlike data that comes from internal sources, most of the big data come from uncontrolled external sources. Therefore, what distinguishes it is its inaccuracy and validity and the lack of clarity of the reliability of the source from which it comes from.

- **The Reliability of the Source:** Data institutions have to make sure that their data sources are reliable before providing false data to its customers. Many social media sites show negative reactions to disgruntled employees about their institutions from which they were released, or to employees critical of competing institutions. These negative statements and comments and, in the absence of verification of their source, may mix with the correct data and provide the producing organization with reliable data that brings benefits and meets the needs of customers.
- **The Amount of Trust to be Shared:** All data may not be disclosed to a wide range of applicants; for example, if the customer interest in the enterprise has data on possible defects in the manufacture of certain products, this interest cannot share this information with all parties that wish to obtain it but must be available only to the manufacturing engineering category that has contracts and charters to the organization.

### 14.3.5 VALUE

Data cannot be judged in advance whether or not it is valuable as raw material, and the concept of value is related to the profit that organizations earn from the use of data.

## 14.4 BIG DATA IN HEALTH

Big data has left a sector only and realized it, and perhaps the most important sector that needs big data analysis in the health sector, which contains thousands of inpatients and outpatients. Each hospital produces about 10 gigabytes of data per year. And to take advantage of this volume, massive data technology has provided many real solutions to medicine, such as developing the treatment of certain diseases, accelerating the launch of effective drugs, and improving healthcare.

Business consultancy McKinsey estimated that the impact of big data applications across the health business could be a reduction of between $300 billion and $450 billion in healthcare spending [8].

### 14.4.1 MEDICAL SEARCH

The purpose of the medical applied search is to improve the health of patients through drug tests, medical equipment, diagnostic methods, treatment protocol, and others. In this context, medical data is necessary to extract statistics in addition to answering the questions asked. The more you get data, the more you get reliable answers [9].

Big data have been used to study many diseases that are increasingly spreading from region to region and from country to country, including epilepsy, cancer, and flu.

One of the companies that used big data in the health field was Google, where at the end of 2008 launched a new app called Google Flu trends. This application works to track the development of influenza disease in real-time, and works on analyzing the words used by millions of users for internet from all over the world in the search engine 'Google' such as cold or flu, or paracetamol, or headache, heat, and another search vocabulary.

Through these words, the application, using a specific algorithm, analyses the disease for each country (29 countries) to extract updated statistics.

To confirm the data provided by the application, Google engineers compared their data with those provided by the U.S. surveillance authorities (Centers for Disease Control and Prevention (CDC)), which was announced in a journal published in February 2009, where the results were close, as well as Showed a close correlation.

The comparison not only showed convergence in data on both sides, but Google Flu Trends was a pioneer in estimating disease indicators. Figure 14.1 shows the comparison between Google data and the U.S. regulatory authority CDC [10].

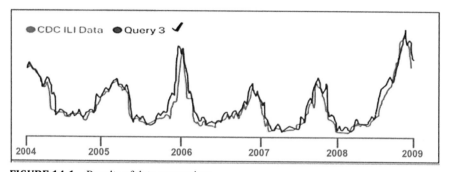

**FIGURE 14.1**   Results of data comparison.
*Source:* Retrieved from: https://www.sciencesetavenir.fr/sante/e-sante/
google-arrete-de-prevoir-mal-les-epidemies-de-grippe_18748.

In 2013, Google provided its data on the disease, noting that there is a difference in the estimates of the spread of the disease between it and the U.S. regulatory authority, where Google estimates exceeded the estimates of the Preventive Authority by 50%. And following these results, many researchers agreed that Google abused itself with its naïve use for big data.

Following this failure to use big data, Google withheld the service from some parties but continued to perform and send its statistics directly to only certain entities, including researchers at Boston Children's Hospital, Columbia University, and Specialists in flue at the U.S. Prevention Agency.

## 14.4.2  PUBLIC HEALTH AND PREVENTION

Big data provides significant assistance to public health, especially in epidemiological surveillance. Monitoring is carried out by analyzing the medical data of patients on the one hand. On the other hand, by analyzing the treatment prescribed to them, so that officials in the public health sector can detect diseases. In addition, the data progress analysis has several development and prevention advantages, resulting in a decrease in the number of infected patients and the cost of their treatment.

In the United States of America, The National Retail Data Monitor: NRDM collects data on the sale of medicines from 29,000 points of sale across the country [10], and the purpose of the analysis is to rapidly detect and prevent epidemics before they spread, expand, and get out of control. The center provides another service to customers seeking information data for specific diseases by customizing a website (Real-time Outbreak and Disease Surveillance) that shows geographical locations that are hotbeds of disease in real-time [11].

## 14.5  HOW BIG DATA IS PROCESSED AT AGENCY FOR HOSPITAL INFORMATION (ATIH) FRANCE

France alone has 260 diverse health databases, with medical, social, and even administrative data for 8 million patients, results of biomedical analyses estimated at 163 million results, and 5 million medical records.

## 14.5.1  TECHNICAL AGENCY FOR HOSPITAL INFORMATION (ATIH)

Technical Agency for Hospital Information (ATIH) was founded in 2000, an institution of a public character (state-owned), whose activity is administrative, operating under the tutelage of the Ministry of Health and the Ministry of Community and Social Security Issues, of the following [11]:

- Collecting, localizing, and analyzing hospital institutions' data regarding their activities, expenditures, organization, and quality of health care, funding, and human resources.
- Technical management of institutional financing mechanisms: calculating annual fees for hospitals and allocating resources.

- Studies on the costs of health and social health institutions: developing curricula and tools, conducting studies, and publishing annual hospitalization costs.
- The agency is also developing and maintaining health codes, such as the international classification of diseases, the standard classification of medical procedures, manuals in the field of occupational rehabilitation.
- The agency is active in the health sector, as it is present in four areas of hospital activity:
  o Medicine, surgery, obstetrics, and dentistry;
  o Home medicine;
  o Follow-up care and rehabilitation; and
  o Psychiatry.

The agency has a multidisciplinary human staff of 122 employees: 34% of statisticians, 25% in informatics, 17% in supporting jobs, 14% doctors, and 10% observers in management and financial analysis.

### 14.5.2   AGENCIES HANDLING OF BIG DATA

The agency conducts a series of processing processes for big data, as the agency uses an information system (inputs, processing, outputs) called the Medical Information Systems Program. PMSI (Information Systems Medicalization Program).

The program works according to the following stages [2, 11]:

1. **Data Coding:** The first stage is considered at the beginning of the operation of the information system when the doctor performs a medical examination in the hospital. He uses codes containing various diseases classified in categories where each disease bears a name and a symbol, each disease has a proper medical procedure for it, and the procedure also carries the name and code to distinguish it from other procedures appropriate to the disease. For the same disease, these symbols are standardized on a global scale and are supervised by the International Organization for Health (OMS). Therefore, the symbols belong to:
   - The patient's illness;
   - The medical procedure performed by the doctor to treat the patient: diagnosis (using a scanner) or therapeutic (surgery).

*Example:* Disease: appendicitis, name of the disease: acute inflammation of the appendix, symbol: K35.8.

*Medical Procedure:* Medical intervention for appendectomy, name of procedure: laparoscopic appendectomy Code: HHFA016.

2. **Collection of Hospital Data:** Each hospital operation leads to the establishment of a set of medical and administrative data, where this data is first recorded on the patient's health record, and then a competent body (Department of Medical Information) at the hospital level extracts medical and administrative data from the register for the patient's health problem, as well as the medical procedures that have been performed to diagnose and treat the patient's condition, the commission then sends this data to the agency ATIH. At the agency level, the nature of the data collected, which will then be stored in codes according to the health code, is determined by the agency, after which the agency categorizes the data into medically and economically consistent groups, so that they are classified according to the nature of health problems (diseases) and medical procedures taken related to it, social, and demographic elements as well as the associated costs. In addition to medical data, the agency collects other data on finance and accounting, human resources, hospital costs, healthcare quality, investments, etc.

3. **Analysis of Hospital Data:** The Agency conducts an analysis of medical data that has already been classified and arranged in groups and uses the analysis extracts by competent and interested health sector authorities in the decision-making process, and the analysis process includes:
   • Analysis of hospital activity: giving a picture of the hospital operation and analyzing its development over time.
   • Financial analysis of health institutions: the aim is to know the financial position of these institutions.
   • Statistics on the costs of hospital activity: determining the costs of each hospital operation and determining the components of these costs.

4. **Recovery of Hospital Information:** Data processed by ATIH can be accessed via the ScanSanté data recovery platform located on the agency's website, ScanSanté allows the service to conduct queries for coordinated information either statistical or health maps, expenses, and analysis of various hospital activities, this information is part of which is open and can be read or uploaded directly by every

researcher, student or interested in medical information, and another part only accessible to actors in the French health system.

5.  **Publishing Medical Information:** The agency provides information on both the activity of health institutions as well as the nature of the diseases they ensure, as the agency distributes the desired information in detail to all bodies that have agreements with the French National Commission for Informatics and Freedom, and the agency has the ability to provide targeted information based on Request from the French Ministry of Health or some national health agencies. Examples of this information include the agency's different studies on young mothers, newborns who are less than normal, than the care process for patients with specific diseases, compared to the hospital accommodation of patients with specific diseases and by geographical region, analysis of absences in health institutions and others.

## 14.6   CONCLUSION, LIMITATION, AND FUTURE RESEARCH

This paper discussed the use of big data in the health sector, which is a burden on the state budget, especially in poor countries, turning it into a sick health sector looking for solutions. In this study, we came up with a set of results that contribute to the development and creation of quality in this sector, which were as follows:

- Open data can bring many positive benefits to society. It promotes the concept of transparency and accountability and promotes a culture of performance, which will bring about development in the health sector.
- Big data technologies and health data analytics provide the means to address the efficiency and quality challenges in the health domain.
- Big data health analytics have the potential to reduce costs of treatment, predict outbreaks of epidemics, avoid preventable diseases, and improve the quality of life in general.
- Big data analysis creates added value, which can be used for economic growth and sustainable development, and stimulates competitiveness.
- Big data helps health institutions become more productive and efficient, like many other industries, adapting healthcare to data analytics not only for their financial returns but also for improving the quality of life of patients.

Through findings, we propose for officials, researchers, and practitioners interested in the health sector, the Algerian, the following:

- Attention to the collection and treatment of big data in the health sector, by linking all health bodies and institutions (pharmacies, hospitals, and clinics of medical analysis and radiology centers.) to digital networks containing different databases, with the purpose of recording big data, and restore them when necessary.
- Training of specialists in big data analysis software and tools.
- Indoctrination of health practitioners, doctors, nurses, pharmacists, supervisors, laboratories, and others, on the importance of big data in the health sector.
- Creating a legal framework governing the use of big data by individuals, institutions, and agencies in order to maintain the confidentiality of data, especially for patients.
- Providing the necessary physical means of high internet, hardware, and software as the necessary infrastructure for the establishment of a big data information system.

Ultimately, any further research in the future could broaden and deepen the debate in order to improve health services and facilities and facilitate the administrative (management) process.

## KEYWORDS

- **Agency ATIH**
- **big data**
- **health**

## REFERENCES

1. https://bigdatainarabic.wordpress.com (accessed on 22 October 2020)
2. Livre Blanc, (2013). *Big Data to Big Business*. https://www.businessdecision-university.com/datas/ck//files/LB-BD-2014.pdf (accessed on 22 October 2020).
3. Krish, K., (2013). *Data Warehousing in the Age of Big Data* (p. 18). Elsevier Inc.; USA.
4. Arvind, S., (2012). *Big Data Analytics* (p. 2). Mc pressed; USA.
5. Rivols, E., (2015). *Data Mining Et Big Data*. http://www.oncorea.com/Presentations/16e%20Rencontre/13.%20Rivals.pdf (accessed on 24 November 2020).

6.  Belouezzane, S., & Ducourtieux, C, (2012). Vertigineux "big data." *Journal Le Monde*.          https://www.lemonde.fr/technologies/article/2012/12/26/vertigineux-big-data_1810213_651865.html (accessed on 22 October 2020).

7.  *Data Transfer Rate*. https://ar.wikipedia.org/wiki (accessed on 22 October 2020).

8.  *Big Data: Can it Make a Real Difference?* http://blog.naseej.com (accessed on 22 October 2020).

9.  Christoph, B., (2013). *Issues and Big Data, Technology, Methods, and Implementation.* https://www.cairn.info/revue-systemes-d-information-et-management.htm (accessed on 22 October 2020).

10. Liu, T.Y., et al., (2013). Association of over-the-counter pharmaceutical sales with influenza-like-illnesses to patient volume in an urgent care setting, *PLoS One Journal, 8*(3), USA.

11. Wagner, M.M., et al., (2004). *National Retail Data Monitor for Public Health Surveillance*. https://www.researchgate.net/publication/8017776 (accessed on 24 November 2020).

# CHAPTER 15

# The Role of Big Data in Avoiding the Banking Default in Algeria (The Possibility of Upgrading the Preventive Centers of the Bank of Algeria as a Source of Big Data)

MOHAMED ILIFI[1] and HAMZA BELGHALEM[2]

[1]*Senior Lecturer, Faculty of Economics, Business, and Management Sciences, University of Djilali Bounaama, Khemis Miliana, Algeria, E-mail: m.ilifi@univ-dbkm.dz*

[2]*Temporary Assistant Professor, Faculty of Economics, Business, and Management Sciences, University of Djilali Bounaama, Khemis Miliana, Algeria, E-mail: hamzabelghalem44@gmail.com*

## ABSTRACT

This chapter aims to demonstrate the role of extensive data in providing the necessary information to the Algerian banks for the effective management of banking risks to achieve safety and security in the Algerian banking units and thus avoid banking stumbling. Stability and centralization of budgets created by the Algerian Bank as a large data source help Algerian banks detect banking risks and avoid banking stumbling. We have noted that these centers have all the huge data attributes, in terms of information volume, velocity, variety, and value, which helps Algerian banks to make better decisions based on the information resulting from the analysis of the huge data through these centers and thus effectively manage risk and avoid bank defaults.

## 15.1   INTRODUCTION

Due to the recent developments in the banking environment, the Algerian banking system has kept pace with its changes and introduced it into the competitive environment.

This has led to an increase in the severity of banking risks to this system, which led to attention to ways and methods that enable it to beat and overcome those risks. However, as banking risks intensified and diversified significantly, this made it difficult to count them due to the high sample size.

Dealing with this type changed the way the statisticians think or even the methods used in analyzing these data quickly and accurately. Furthermore, the enormous amount of data produced, processed, and made accessible to banking institutions has become a source of strength for the accurate and fastest possible identification of risks. Therefore, Algerian banks should adopt preventive centers to effectively store and analyze information, particularly after the issuance of monetary law, which helps banks avoid bank defaults.

To take note of the various aspects of this topic, we divided this chapter into three main themes:

- **Section 15.2:** Theoretical framework for big data.
- **Section 15.3:** Basic concepts of bank default.
- **Section 15.4:** The role of preventive centers as a source of big data in avoiding the problem of banking default in Algeria.

## 15.2   THE THEORETICAL FRAMEWORK OF BIG DATA

In fact, no clear definition of big data can be given, as it is a complex, polymorphic term, whose definition differs between the communities you are interested in as a user or service distributor, etc. It is called big data, massive data, and megadata.

### 15.2.1   THE CONCEPT OF BIG DATA

Big data is defined as the balances of information that are large in size, speed, and diversity that require innovative and effective forms of processing that are different from normal data processing so that their users can improve visibility [1], decision making, and automation.

Data, etc., cannot be efficiently processed using existing and traditional technology to benefit from them. In addition, versatile compared to types of commonly used data sets [2]. According to the previous definitions, big data can be defined as a set of new mechanisms used to analyze and draw conclusions of value in the analysis.

## 15.2.2   CHARACTERISTICS OF BIG DATA

In order to talk about big data, it must provide four characteristics mentioned in the following [3]:

- **Volume:** Refers to the amount of data that grows day by day at a very rapid pace, the volume of data generated by humans and machines, and their interaction on social media.
- **Velocity:** Speed is defined as the frequency at which different sources generate data every day.
- **Variety:** Since there are many sources that contribute to big data, the type of data it generates is different, and it can be regular and irregular, and therefore there is a variety of data from the data generated every day.
- **Value:** After discussing all the characteristics above, there is one to be taken into account when looking at big data, which is the value, everything is good in accessing the data, but unless we can convert it to a value, there is no use.

## 15.2.3   THE BENEFITS OF USING BIG DATA

The benefits of big data analysis in an organization of any kind lie in the following points [1]:

- Identifying the faults and weaknesses and improving operations in all financial units and management;
- Making better decisions based on information generated by big data analysis for all financial and administrative units;
- discovering untapped opportunities and potential weaknesses in all actions;

- Enabling stakeholders to find solutions to potential problems encountered in the analysis of big data in certain processes;
- Increased opportunity for clear and correct decisions.

## 15.3 BASIC CONCEPTS OF BANKING DEFAULT

The failure of banks has become a problem afflicted by many banking systems in many countries of the world, resulting in huge financial, economic, and social losses, and this problem has emerged as a result of the unconscious expansion of the banking industry activity internally and externally in order to attract more customers and achieve returns.

### 15.3.1 DEFINITION OF BANK FALTERING

Projects are usually seen in two ways [4]:

1. **Financially:** An enterprise is faced with unforeseen (emergency) circumstances that lead to its inability to generate an economic return or a cash surplus that is not sufficient to meet its short-term obligations and the inability to cover them from external sources;
2. **Economically:** The faltering project is where the rate of return on investment below the cost of capital and thus characterized by its inability to meet the obligations owed despite the size of assets to its liabilities. It is also perceived that the establishment's failure to meet its accumulated financial obligations towards its creditors may lead to a partial or total cessation of the activity, which threatens its legal existence itself, i.e., entering the bankruptcy stage. This is a condition under which a commercial bank is unable to satisfy its obligations or meet the customer's demand for liquidity in the short term due to its lack of liquidity, its problems and shocks, the worsening of bond prices, and the bank's poor governance, etc.

### 15.3.2 TYPES OF DEFAULT RELATED TO BANKING ACTIVITY

The type of default facing banks is related to the set of circumstances related to banking activity at the macro level, as well as the nature of internal factors

related to the banking institution, so the default can be divided into the following types of banking [5]:

1.  **The Default of the Banking Activity:** It is also expressed by the failure related to competition, which refers to the situation in which the bank is unable to compete in the banking market at both its local and international levels, which translates into a decrease in its market share (low returns, low deposits, and increased liabilities). This means not being unable to continue and exit the banking market.
2.  **The Default of the Banking Institution:** This form is situated at the level of the bank itself and involves a detailed review of the budget for the financial year, the estimation of the results, the identification of the default size, and the most important factors that have contributed to its development. Depending on the severity of the default, the collapse of a financial institution takes several forms:

    - **The Default:** It related to shareholders: the bank's position is characterized by the adequacy of its assets to cover the liabilities of third parties in all its variations without the liabilities of the shareholders;
    - **Excess Default:** The bank is unable to meet the obligations payable from its available liquid assets, although the actual valuation of its total assets covers its liabilities and liabilities to third parties. The bank is in a state of technical, financial difficulty (temporary liquidity problem), but it is evident (apparent) that may present the bank to declare a suspension of payment if it fails to take the necessary measures to secure the necessary liquidity;
    - **Short-Term Capital Flows:** Where the bank's assets become much less than the liabilities of others, providing adequate liquidity to repay the part is needed. However, if all procedures lack the sources of liquidity, the default is characterized by being exposed and in real financial hardship (bankruptcy).

### 15.3.3   THE CAUSES OF BANKING DEFAULT

The causes of banking default are divided into internal and external causes [6]:

1. **External Causes of Banking Default:** Reflect the reasons that originate from the general external environment, which do not fall within the bank's scope of control. The most important of these reasons are:

   - **Macroeconomic Instability:** It is the imbalance caused by successive changes in the structure of the national economy, such as fluctuations in terms of trade. When low trade conditions, it is difficult for bank customers engaged in activities related to export and import to fulfill their obligations, especially debt service, as well as fluctuations in the inflation rate. Critical in the ability of the banking system to act as a mediator, especially credit, liquidity, and other macroeconomic fluctuations.

   - **Short-Term Capital Flows:** These flows increase the amount of bank deposits that increase credit irrespective of the solvency of the beneficiaries, the accumulation of bad debts and increasing bank losses, and the increase in the number of cases of defaults, as shown during the Asian crisis in 1997. Inadequate economic and financial reforms: especially exaggerated, was one of the main reasons behind the default of state banks in the 1980s.

   - **Government Intervention:** Some governments use their banks to finance ventures that are profitable and are seen as the key public treasury financed through lending to the public sector at high rates of their own and non-self-resources, resulting in major difficulties and problems due to the poor performance of financed firms.

   - **Legislative Reasons:** The laws governing banks' business are insufficient in their coverage and coverage of many of the gaps that appear in banking, which do not help in taking the necessary measures in a timely manner.

   - **Regulatory Factors:** The absence of effective banking supervision and supervision and the adoption of a system of fixed deposit insurance."

   - **The Risk of Intensive Withdrawal of Deposits:** Known as banking panic, a phenomenon that occurs due to the emergence of the problem of the failure of a bank, where creditors and depositors withdraw their dues in a sudden and collective manner.

   - **Bank Policy Error Should Not Fail:** This policy is based on the fundamental belief that the bankruptcy of a large bank has

wide negative financial effects, especially the transmission of financial contagion (bank panic) to other banks, especially small ones, and therefore should prevent the collapse of this bank. However, this policy gives the wrong signal to the big bank after helping it (not forgetting the other big banks), to engage in high-risk financial and banking activities, exposing it to defaulting situations again.

2.  **Internal Causes of Banking Default:** Internal causes of bank default include the majority of reasons that the bank can control because it is due to the bank's internal environment. The most important of these reasons are:

- Poor management and low efficiency of administrative competencies.
- Fraud and Corruption: Ethical causes of banking default are among the most important challenges facing banking supervision, as it is difficult for the supervisor to detect them due to the tightness in their planning and implementation, and the fraud that is exposed to the bank stems from employees or customers, including internal lending to bank managers. In addition, many of the banks that faced the problem of default were using fake accounts and creating false and false names of borrowers, so that they could escape the control of the monetary authorities. In the previous occurrence of default in cases where management control weakens the performance within the bank and the absence of controls, which monitors their performance and, in particular, the performance of basic and influential administrative elements within the bank and the focus of the decision within one person, etc.
- Marketing reasons: Ultimately, the absence of an efficient and strong marketing device within the bank leads to the inability to face changes in the banking industry, both domestic and global, most studies in this area found out that the key reasons for bankruptcy are the deficiency of marketing skills and the lack of choice of the right place for marketing.
- Increasing the volume of non-performing loans;
- Lack of bank liquidity;
- Insufficient capital.

## 15.3.4  EFFECTS AND MANIFESTATIONS OF BANKING DEFAULT

According to the experiences of countries affected by the problem of banking faltering, the main negative effects of this problem are summarized as follows [7, 8]:

- Slowing economic growth;
- The vibration of confidence in banks;
- Transformation of short-term cash flows;
- Affected by the relations of the local banking system;
- The default of projects and companies financed by banks.

The manifestations of faltering are divided into internal and external aspects that will be exposed to them as follows [9]:

1. **Internal Manifestations:** They are summarized in the following points:

   - The volatility of financial ratios fluctuated in successive periods, mainly liquidity, profitability, and asset quality.
   - Low operating profit and deterioration for successive periods and the prospect of continuing for years to come due to the size of the losses inflicted on the bank. Thus, the erosion of capital and equity management conflicts internally under the critical situation of the bank.
   - The low morale of employees and their tendency to leave the troubled bank, especially if confirmed liquidation in the future.
   - The imbalance of the bank's financial structure, such as the increasing reliance on borrowing (the high due to banks).

2. **External Manifestations:** These aspects relate to the external environment of the bank, which we will highlight the following:

   - The high volume of deposit withdrawals by depositors, especially in the case of the announcement of the troubled situation of the bank or because of information and rumors and increases withdrawals if there is no system of deposit insurance.
   - The inability of the default bank to meet its short-term obligations on the due dates.
   - A sharp and apparent decline in the market value of the bank's listed and listed shares.

- The reluctance of correspondent banks dealing with the bank to complete operations related to this type of activity.

## 15.4   THE ROLE OF PREVENTIVE CENTERS AS A SOURCE OF BIG DATA IN AVOIDING THE PROBLEM OF BANKING DEFAULT IN ALGERIA

The Algerian legislator has supported banking supervision with other preventive centers, especially in the area of credit risk.

### 15.4.1   CENTRALIZATION OF RISKS

This department was established under Article 106 of Law 90–10 on Cash and Loan, which is called the Risk Center. Currently, it is regulated and operated by Regulation No. 12–01 of 20 February 2012, which regulates the centralization and operation of the enterprise and household risks, which is called the centralization of risks. According to the latter, the centralization of risk is divided into two [10]:

1.  **Centralized Enterprise Risk:** Data on loans to legal and natural persons are recorded unpaid professional activity.
2.  **Centralized Household Risk:** Collects data on loans to individuals. Risk centralization is a risk centralization service charged with collecting, in particular, each bank and financial institution (called authorized institutions), the identity of the beneficiaries of the loans, the nature, and ceiling of the loans granted, the amount of uses, the amount of outstanding loans and the collateral is taken for each class of loans. Authorized institutions in this regard shall be authorized to centralize risks according to the nature of the data in their institutional section and in their household section [10]:

    - Data on the definition of the beneficiaries of loans and the ceiling of loans granted to customers, whatever the amount, with the title of transactions carried out at the level of their windows as well as guarantees taken in kind or in-person for each type of loan, this information is called positive data;
    - Unpaid amounts from loan lists; this information is called negative data.

Algeria's centralized risk department is one of the models that are centralized and managed by central banks (government agencies). These systems are characterized by the obligation to provide banking institutions with credit data and information, thus ensuring greater responsiveness by these institutions, in addition to the supervisory authority. In dealing with this data and knowledge, credibility, and accountability lead to greater assurance of the integrity and good use of this data. In addition, the key drawbacks are the data obtained in it. They are all negative facts about default, bankruptcy, and filter events, and do not provide the positive information that Order 12-01 has an impact on credit-building [11].

In order to activate the role of this department, the Algerian Bank established from 2005 to 2009 a system of online consultation, which was usable since 2006. By the end of 2010, this system became centralized. This was after the file contains 69657 permits in 2009, which indicates the positive development of the number of permits submitted by banks and financial institutions; for further clarification, see Table 15.1.

**TABLE 15.1** The Number of Permits for Algerian Banks and Financial Institutions to Centralize Risk for 2011–2017 has Evolved

| 2017 | 2016 | 2015 | 2014 | 2013 | 2012 | 2011 | Year |
|------|------|------|------|------|------|------|------|
| 584,807 | 719,722 | 412,147 | 35,006 | 43,584 | 4,510,599 | 3,537,951 | Number of permits |

*Source:* Bank of Algeria [12].

Table 15.1 shows that the number of permits is constantly evolving, reaching 719,722 permits in 2016, compared to 3,537,951 permits in 2011. This is a positive indicator to support the process of reducing credit risk in Algerian banks by avoiding dealing with doubtful borrowers. This allows for avoiding the problem of banking default from this angle.

### 15.4.2 CENTRALIZATION OF PAYMENT GIRDERS

The Bank of Algeria's Reserve and Prevention Bank, more than the risks associated with banking operations, established girders' centralization under order No. 92–02 of 22 March. This centralization for each payment instrument and/or loan shall [11]:

- To organize and maintain a central index of payment impediments and consequent follow-up.

- Periodically notify the financial intermediaries and each other designated authority of the list of payment impediments and consequent follow-up.

This centralization was supported by the issuance of Regulation No. 08–01 of 20 January 2008 on arrangements for the prevention and control of the issuance of unencumbered checks.

Algeria, for the purpose of reviewing and exploiting it, especially when delivering the first checkbook to its customer, and once a payment bidder has occurred due to lack or lack of balance, the drawee shall, in accordance with the provisions of the commercial code, authorize the centralization of payment girders within four working days following.

It shall also require the drawee to send to the issuer of the check within the stipulated time a directive stating that the bidder has been authorized to centralize the payment symptoms. And in case of non-settlement within 10 days, it will prevent the issuance of checks within a period of 5 years to all authorized institutions as of the date of the order instructed and initiate criminal proceedings in the absence of settlement [13].

It is clear from Table 15.2 that the number of permits for centralization of payment beams witnessed a fluctuating development during the period of analysis, but from 2011 it increased to 2016 with a decrease in 2017, which means high risks for banks and financial institutions.

**TABLE 15.2**  The Development of Permits in the Centrality of Payment Symptoms During the Period 2011–2017

| 2017 | 2016 | 2015 | 2014 | 2013 | 2012 | 2011 | Year |
|------|------|------|------|------|------|------|------|
| 50713 | 65263 | 62267 | 56572 | 4458 | 44207 | 43266 | The Issue of Dud Checks |

*Source:* Bank of Algeria [12].

Issuing Regulation 11–07 of 19 October 2011, amending, and supplementing Regulation No. 08–01, which, in the event of a bidder, paid the preparation and delivery of a non-payment certificate to the beneficiary of the outstanding check not paid by the bank submitting the check to the rejection of the check at the electronic clearing system [13].

## 15.5  CONCLUSION, LIMITATION, AND FUTURE RESEARCH

The study of the role of big data in avoiding the failure of banking units in Algeria has also benefited us with the following results:

- Big data refers to the ability to draw insights and make decisions, using a set of modern technology tools, which store and process data and translate it into information to use regulators in conducting their banking activities.
- Big data in banking institutions aims to: identify weaknesses and weaknesses and improve operations in all financial and management units; make better decisions based on information generated by analyzing big data for all financial and administrative units; discover untapped opportunities and potential weaknesses in all businesses; concerned solutions can be found to detect when analyzing big data potential problems in some processes; and increase the chance of making clear and correct decisions.
- The problem of bank default is defined by the inability of the bank's capital to meet short-term obligations to depositors.
- The effects of the banking default are all due to low confidence in banks, low economic growth rates, and the deterioration of the relationship between the domestic banking system and international institutions, especially in the field of finance.
- The preventive center, which is represented by the centralization of risks, the centralization of payment gateways, and the Commission on Banking Stability, on the early detection of the risks surrounding the banking institution, and on the weaknesses that it knows, through its analysis of the indicators and extracts information, and then takes measures. In view of the above results, we make a set of the following recommendations:

  - Establishment of laboratories specialized in data analysis within the preventive centers adopted by you, Algerian.
  - Periodic training courses for workers at the level of banking institutions, with a view to updating information and introducing innovative analysis methods.
  - Regulate the provision of big data services by working on legislation to regulate this service to overcome these challenges.
  - Increase in the composition of specialized frameworks for analyzing big data by banks' responsibility to respond to the banking risks in the contemporary banking environment on time.

## KEYWORDS

- **banking stumbling**
- **big data**
- **preventive centers**

## REFERENCES

1. Abdul, R. M., & Rashwan, S., (2018). The role of big data analysis in the rationalization of financial and administrative decision making in Palestinian universities: A field study. *Economics and Finance, 11*(01), 27-29.
2. Organization International De Normalization, (S.D.). *Learn About Big Data* (p. 1). www.Inoledge.Com/Post.Php?Id =91 (accessed on 22 October 2020).
3. Nahas, A., (2017). pp. 8, 9. Shamra.Sy/Uploads/Documents/Document_ A83f7d9e9bf884b 6d5ee20e15dddb4ff.Pdf (accessed on 22 October 2020).
4. Al-Zaher, M., & Al, E., (2007). Factors determining the failure of banking facilities in Palestinian banks. *An-Najah University Journal for Research, 21*(02), 518, 519.
5. Chambour, T., (1992). *The Tumbling of the Banking Corporation in Lebanon, Research and Discussions Symposium on Non-Performing Banks and Means of Treatment* (pp. 33–36). Union of Arab Banks.
6. Ilifi, M., (2013–2014). *Methods of Minimizing the Risk of Banking Defaults in Developing Countries with the Case Study of Algeria, Doctorate in Economic Sciences* (pp. 78–82). Hassiba Ben Bouali University of Chlef.
7. Fattah, A., (1992). Bank default and its remedies the case of Jordan. *Research and Discussions Symposium of Troubled Banks and Means of Treatment* (pp. 223–225). Union of Arab Banks.
8. Al-Khazali, A. S. K., (2000). *Banking Defaults in Jordan: A Comparative Analytical Study (1980–1998)* (pp. 82–84). Master in finance and banking, faculty of economics and administrative sciences, Al-Bayt University.
9. Fareed, H. M. W., (2008). *Building a Model for Predicting the Financial Failure of Jordanian Public Shareholding Companies Working in the Insurance and Banking Sectors, Doctoral Dissertation* (pp. 18–20). Oman, Graduate School of Administrative and Financial Studies.
10. Bank of Algeria, (2012). Regulation 12-01 of 20 February organization and running of the credit risk registry of companies and households. *Algeria: Official Journal of the Republic of Algeria*, 45.
11. Bernie, M. I., (2019). *Developing Credit Information System and Risk Conventions Arab Countries* (pp. 17, 18). www.Amf.Org.Ae/Sites/Default/Files/Econ/ Amdb/%5Bvocab%5D/%5Bterm%5D/2.Pdf (accessed on 22 October 2020).
12. Bank of Algeria, (2011–2017). Report on Economic and Monetary Development in Algeria.

13. Bank of Algeria, (2008). Regulation 08–01 of 20 January 2008 Relating to the system of prevention and fight against the issuing of dud cheques. *Algeria: Official Journal of the Republic of Algeria,* 21, 22.

# CHAPTER 16

# Marketing Information System as a Marketing Crisis Management Mechanism Through Big Data Analytics: A Case Study of Algeria Telecom in Bouira

RABAH GHAZI and FATIMA ZOHRA SOUKEUR

*Laboratory of Globalization, Politics, and Economics, University of Algiers 3, Dely Brahim, Algeria, E-mails: ghazi.rabah@univ-alger3.dz (R. Ghazi), zola_marketing@yahoo.fr (F. Z. Soukeur)*

## ABSTRACT

This chapter aims to address the marketing information system MIS and its role in the face of marketing crises, especially in light of the ongoing changes and what needs him to current information about the surrounding environment. The study concluded that the marketing information system MIS plays an active role in the management of the marketing crisis by properly exploiting the huge amount provided by big data of classified data, and stored according to a database that helps to speed the flow to allow for the response of these crises and predict.

## 16.1 INTRODUCTION

This suggests how important it is for organizations to have an information system to deal with the marketing crises they encounter in the face of constant changes. Therefore, marketing crisis management needs to present information about the environment in which they operate. Based on this, one can ask the fundamental question of this study:

*How much does the marketing information system MIS in the management of marketing crises?*

To answer this question, this study is divided into the following axes:

1. Conceptual framework for big data, marketing information system, and marketing crisis management.
2. Marketing information system in the administration of the crises turned marketing.
3. A field study on the Algerian telecom corporation in Bouira.

We believe that this study is of great importance in the life of the organization can be mentioned in:

- Highlighting big data, which has become a haunting issue, has become the focus of many researchers and decision-makers in government sectors and companies more.
- Given the ongoing developments and rapid shifts in the field of information technology, where data has become available here and there, and how well the marketing information system can benefit from it.
- The great importance of marketing crises that have become a real concern for managers of organizations in general and marketing officials in particular.
- Highlight the role played by the marketing information system in collecting information from big data quality, which helps to make effective decision-making.

This study also aims to:

- Attempt to highlight the value of big data as a modern trend to provide the MIS with the right data at the right time.
- Try to highlight how the marketing information system MIS contributes to the management of marketing crises.
- To clarify the importance of MIS marketing as a mechanism for combining the usable outputs of big data to address marketing crises.
- Raising awareness among organizations of the need to pay attention to MIS's marketing information system in responding to crises.

## 16.2   CONCEPTUAL FRAMEWORK FOR BIG DATA, MARKETING INFORMATION SYSTEM, AND MARKETING CRISIS MANAGEMENT

Organizations that have big data face the challenge of being able to control them. The best way to store, manage, and use this data is a real opportunity, and analyze them because they provide a deeper understanding of the people and things about which data was formed in the organization, and this helps.

Officials make the right decisions to give them a future vision in the face of these crises. Big data has become one of the most popular terms a popular terminology in circulation recently. Everyone is talking about it all speaks about a final.

What is big data? Watson in Ref. [1] claims that "It is a large amount of data and a variety of disorganized, which makes dealing with it is very difficult." It is defined as "data that is beyond the ability of ordinary databases to process, which is very large data moving at high speed does not meet the requirements of the structure of your database and to benefit from this data, it is better to choose the most appropriate alternative for processing."

From these definitions, we conclude that big data has a set of characteristics:

- Very large size;
- Diversity in the source, scope, and format;
- Speed of production;
- The added value of this data when analyzed and processed.

Hence, the major importance of big data is that it is possible to improve efficiency in the context of in the course of using a large amount of data of various types. If the big data is correctly defined and used accordingly, organizations enable to have a better view of their business and thus lead to efficiency in different fields. For example, sales, factory product improvement, service provided, etc., use big data effectively in each of the following areas [2]:

- In information technology: to improve security and troubleshooting by analyzing patterns in existing records;
- In customer service: using information from call centers in order to obtain customer style and thus enhance customer satisfaction by customizing services according to their orientation and desires;

- In improving services and products: Through social use of content methods of the continuity and defined knowledge of possible preferences the agents of which the organization from modification be possible her expected products and avoidance of the crises;
- In detecting fraud in online transactions for any industry;
- In reducing risk: by analyzing the information of transactions in the financial market.

Big data also vary in terms of sources and access methods. ECE categorizes big data sources as follows [3]:

- Data originates from transactions between so replies, such as credit card transactions and online transactions, including through mobile devices.
- Data captured via sensor networks, such as satellite imaging, road sensors, climate sensors, etc.
- Sources received through tracking devices, such as tracking data from mobile phones and GPS.
- Track behavioral data, such as how often you search the Internet for a product or service or another type of information, and how often a particular segment is viewed online.
- Surveys data, such as social media comments.

If, in light of this huge amount of data and for the purpose of marketing them, the institutions seek to exploit them through tight systems. Perhaps the marketing information system is the way to do so. Are support and a pillar to make those decisions? What is the marketing information system?

Kotler defines him as: "The complex and integrated structure of human frameworks, devices, and procedures, which is designed to collect data from internal and external sources of the facility to generate information that helps marketing management in making sound decisions" [4].

It is also defined as: "A structured way of continuous collection, evaluation, analysis, storage, and transfer of information to marketing decision-makers" [5].

A set of human and mechanical elements is necessary to collect data to convert it into information that helps management make accurate and successful marketing decisions.

By addressing these concepts, the following definition of a marketing information system can be derived as "an interconnected group of people, machines, and software that collect, analyze, and store data from their various sources, in a regular manner for their timely use."

Through this, we see that marketing information systems have several characteristics that can be listed as follows [6]:

- It is the application of the concept of systems in the field of information through (identification of required data, their collection, operation, and storage);
- It is concerned with the future: it anticipates, prevents, and solves problems; it is both therapeutic and preventive;
- Continuous: not related to the study of a problem but working in an ongoing manner;
- Consider wasteful and expensive unless the information provided is used.

It also provides significant importance and benefits to the organization in general and marketing managers in particular because of it [7]:

- Determine to the management the path to be followed in the trade-off among the alternatives available for marketing decision-making.
- The work of the institution is perceived as all integrated rather than separate parts and integrated analysis.
- Enables enterprise systems to automatically retrieve a huge body of information, calculate each transaction's share, each commodity, and everyone in the organization's profits, thereby identifying management trends and policies.
- The ability to modify information effortlessly and answer any questions related to its customers and products immediately as used in the analysis of the institution's daily activity.
- Assists in the use of electronic computers, computer programs, and modern methods of communication and benefit from them in the field of information.

On the other hand, business organizations face different crises in different stages of their life cycle.

The marketing crisis is defined as the "decline or decline of marketing's strategic role within organizations." This was agreed upon by the marketing academy in the early 1980s [8].

It is also known as "weak or rigid product discharge, low market share, and failure to cover all markets in a given period of time" [9].

Through these different definitions of marketing crises, we note that they have a set of characteristics, which are summarized as follows:

1. **The Importance of Betting:** During an institution's crisis, it is in a state of pressure, and if the rate of achievement of its objectives has weakened the scope of the crisis, the latter is deemed to be effective only if that objective is of critical importance to the institution.
2. **Risk of Loss:** There are three factors that allow the assessment of the risk faced by the institution. The first is manifested by the confidence it gives to the process of setting goals and determining the current situation. If there is a possible solution to the form at hand, the risk is weak, and the crisis decreases.
3. **Sudden:** Some sectors such as the transport sector and the chemical industry sector are at constant risk due to the nature of their activities; therefore, we find an index of all accidents, but others exposed to sudden or unexpected crises denies the preparation of a plan to ease the crisis consequences.

In order to give more insight into the marketing crises, it is necessary to address the types of the latter, which are classified as [10]:

- **By Domain:** According to this criterion, we find that the marketing crises are multiple according to the scope of its activity, strategic marketing, social marketing, etc.
- **By Sector of Activity:** There are commodity marketing crises (consumer or industrial), or marketing crises of services (telecommunications, tourism services, health services, etc.).
- **By Geographical Level:** Marketing crises can arise at the enterprise, regional, or national level and may extend beyond geographical space.
- **By Recurrence:** The crisis occurs depending on the recurrence of factors of origin and causes.
- **By Topic:** There are many topics of the marketing crisis that may be the theme of the brand; maybe consumer tastes or may be related to the policy of marketing mix, etc.
- **According to its Stage:** During the stages of the organization's growth, there are several marketing crises that can be classified according to its life cycle (presentation stage, growth, maturity, deterioration).

The concept of marketing crisis management refers to how to overcome the crisis using a method in order to minimize it's negative and maximize its positives and is defined as [11]:

- A meaningful activity based on research and obtaining the necessary information to enable the administration to predict the places and trends of the expected crisis, and to create the appropriate climate to deal with it by taking measures to control and eliminate the expected crisis or change course for the benefit of the institution;
- As defined by Abu Qahf as "preparations and administrative efforts to face and reduce the destruction resulting from the crisis."

Thus, the definitions addressed to us shows that the management of marketing crises go through the following stages [12]:

1. **Early Warning Phase:** It is a warning phase to recognize the crisis and is the signals that threaten the occurrence of a crisis, and if not realized, the crisis phase comes quickly, and maybe direct warning can be perceived and maybe the opposite, and here the organization is trying to discover the causes and variables that lead to the occurrence the crisis.
2. **Crisis Burst:** When the organization fails to act at the previous stage (crisis) or fails to make the appropriate decision, or it cannot control the variables of the accelerating situation, the organization is exposed to a crisis with a high degree of strength, intensity, and violence.
3. **Phase of the Decline of the Crisis:** The stage of crisis resolution and management is to control the crisis and get rid of it after achieving its results.

## 16.3   EXPERIMENTAL METHODS AND MATERIALS

### 16.3.1   THE ROLE OF MIS MARKETING INFORMATION SYSTEM IN THE PREVENTION OF MARKETING CRISES

The current era is witnessing a great digital revolution in which information has become a two-edged sword:

- Treatment and analysis of big data for organizations enable them to reach very advanced levels of predictive intelligence and understand the pattern, behaviors, and desires of the people they deal with.

- It is also worth noting that big data contains many types of data that can be used to a great extent, such as images, audio, videos, 3D models, and electronic navigation maps data; ignoring or postponing big data is no longer an option and does not achieve any relevant result. Organizations have to work on acquiring data analysis and processing systems and tools so that they can extract their components from very useful information.

- Organizations are now acquiring additional data from their operating environment with increasing speed, namely:

  o **Web Data:** Web-based customer behavior data can be captured, such as page views, searches, read comments, and purchases. It can enhance performance in best future views, model laying, customer segmentation, and targeted advertising.

  o **Text Data:** e-mail, news, Facebook publications, documents, etc., (One many of the absolute kinds of the data on a full scale, where the focus is usually on extracting key facts from the text and then automatically using insurance as fraudulent or not).

  o **Time and Location Data:** GPS and mobile technology, as well as Wi-Fi, make time and location information a growing source of data and at the individual level. With the amount of data generated, several interesting apps are beginning to appear, and their privacy should be managed with great caution.

  o **Smart Grid and Sensor Data:** Smart Grid and Sensor Data are now collected from automobiles, oil pipelines, and windmill turbines, and are collected at an extremely high frequency. Just as sensor data provides robust information on engines and machines' performance, problems can be diagnosed more easily and faster to develop procedures.

  o **Social Network Data:** Within social networking sites such as Facebook, Twitter, and Instagram, it is possible to do a link analysis to detect a network from a particular user. Social network research may offer ideas about advertising that can attract particular users. This is done not only because of the desires individually listed by consumers, but also because they see what is important in their circle of friends or colleagues (Figure 16.1).

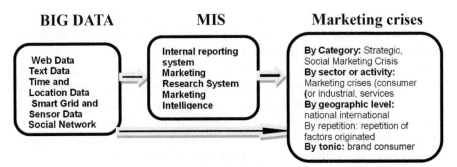

**FIGURE 16.1**   The role of MIS in the management of marketing crises through big data.

The marketing information system performs a set of related functions among them; starting with collecting, sorting, operating, and analyzing big data, saving the indicators and marketing information extracted from them, and retrieving them when needed to provide decision-makers with the right time and the right quantity and type as well.

Computer software and analysis software to minimize the probability of error and thus predict the crisis before it occurs. The marketing information system provides the crisis management team with various marketing studies and research, which enables it to monitor early warning signals that may occur in the organization's marketing activity and thus improve its performance, hence its vital role in the prevention of marketing crises.

Information also plays an important role in streamlining the organization's marketing mix decisions by collecting crisis information to analyze product performance to preserve and protect the organization's products from marketing crises, whether new products are introduced, deleted, or improved.

The marketing information system also provides the crisis management team with information on competing and alternative commodities and information on the degree of elasticity of demand for FAO products, prices of competing products, and general economic conditions to monitor early warning of pricing or re-adjusting current prices.

It also provides information on the nature of the market and the characteristics of the advertising message, and the positioning of the product from its life cycle to contribute to the identification of appropriate promotional methods and the definition of promotional objectives, enabling it to prevent the occurrence of the crisis.

Therefore, the role of the marketing information system is also to give a picture of the nature of the markets dealt with by the organization to choose the distribution outlets by providing information.

The marketing information system relies on the use of technological devices that build, continue to update, and analyze basic databases and provide them with the marketing crisis management team in a timely manner to prevent the marketing crisis. The continued flow of information during the marketing crisis management phases is a key factor in decision-making speed in response to the crisis and its possible repercussions.

Confirm the information after sorted by the marketing crisis management team, prove its safety and relevance; store it for exploitation for vigilance, and monitor early warning signs of a marketing crisis.

### 16.3.2   THE FIELD STUDY ON ALGERIEN TELECOM IN BOUIRA

The following will attempt to diagnose the marketing information system at the level of Algerie Telecom under study:

> **The Accounting System and the Internal Records System:** The accounting system at Algerie Telecom is based on a set of records by which the accounting data is processed:

- The stage of recording information from documents.
- The processing phase, where the account or group of accounts where the information is recorded, is identified.
- The stage of storage. Here, at this stage, legal procedures must be respected, in particular, Article 09 of the Algerian Commercial Code.
- The stage of providing information to the Directorate General and other structures in the form of summary tables such as the table of accounts of the budget results.
- The accounting system is an important source within the organization to provide information that helps in the decision-making process, drawing plans, strategies, and expectations.
- The accounting system allows the collection, processing, and communication of information necessary for the purpose of understanding the activities of the institution and determine the level of performance. Economic activity in aggregate and clarification of profit and loss in the relevant financial period takes certain figures to distribute profits and pay taxes. In addition,

within the complex, the Foundation uses analytical accounting to allocate and charge each interest with its associated costs. In other words, the purpose of accounting is to produce analytical results to help managers make appropriate decisions to emerge from the crisis.

### 16.3.3   MARKETING RESEARCH SYSTEM

Despite the great importance of the marketing research and studies system as well as marketing intelligence and despite the importance of the institution that is active in the field of communications.

But what we observed through the study in the field of research and marketing studies that enable the collection and presentation of information about the prevailing market conditions, the situation of competitors, demand trends, etc., and their contribution in predicting some marketing problems that they may encounter and work to solve them.

We need to study the markets and consumers' behavior in light of the openness of the Algerian economy, which creates a more competitive environmental force than before, and therefore organizations that do not adopt this type of access to information cannot continue their activities and achieve their objectives.

Marketing research is not an end in itself, but a means of providing data and information that will help diagnose and find solutions to marketing crises.

### 16.3.4   MARKETING INTELLIGENCE

While the marketing intelligence system relies on observation to capture and capture information and access to journals and studies of specialized offices and university research, Algeria Telecom depends on exploiting the opportunity of missions sent when concluding contracts with foreigners.

Participation in national and international exhibitions would provide a general overview of world markets as well as the way of advertising and services, and participation in an international economic magazine specializing in telecommunications topics, which helps them to deal with crises related to promotional activities or so-called misleading advertisements that leave a bad mental impression the customer.

The marketing decision support system is defined in the processing programs included in computers and intranets adopted by the institution to exchange information between its various units.

The marketing decision support system is the result of the widespread use of computers in various marketing activities, and the multiplicity and diversity of variables, problems, and crises faced by the Algerian Telecom Corporation, which makes it difficult for the human mind to conduct an accurate and rapid analysis of the many relationships between these variables and their mutual impact.

In light of the above diagnosis of all parts of the marketing information system in the Algerian Telecom Corporation can be reached the following results:

- The effectiveness of this system leads to the ability of the institution to diagnose its marketing environment, and thus, knowledge of the various external marketing variables and crises that may be encountered.
- The evolution of this system is considered to be the institution's progress in ensuring a means to ensure the flow of information, especially those coming from.

The external environment, which has a significant impact on marketing decisions and the design of marketing strategies of the enterprise:

- The effectiveness of this system refers to the extent to which the institution is aware of the task of strategic vigilance, which has become an activity adopted by various organizations that want to stay and lead in a competitive environment and provide correct and meaningful information.
- Exchange of information between the institution's various administrative systems to serve as a coherent chain of familiarity with all the problems and crises in each department to prevent and predict them.
- Placing sufficient information at the disposal of officials to answer customer questions and study their needs.

## 16.4   CONCLUSION

The exit of the organization from the marketing crisis depends on the proportion of information confirmed who has accurate information has the ability to make the most difficult decisions, but if the information is incomplete or incorrect, the degree of risk and uncertainty is big, and therefore cannot monitor early warning signals and difficulty of predicting a marketing crisis.

Therefore, the marketing crisis prevention process requires that marketing information be accurate, fast-flowing, clear, and the need to follow the feedback of data and work on tabulation, and stored using electronic devices according to a database that helps to speed the flow.

> **Results of the Study:**

- MIS marketing information system is an effective means to assist decision-makers in the face of possible crises on marketing performance.
- Big data outputs are inputs to the marketing information system, and marketing information system outputs serve as the organization's reliable database for marketing crisis management.
- The marketing information system collects and analyzes the various data of marketing activities for use in resolving marketing crises by predicting them before they occur and thus preventing them.
- The crisis is a point of emaciation in the life of the institution, both for the worse or for the better.

> **Limitation and Future Research:**

- analyzing big data for its untapped treasures and benefits, from which undisclosed results are drawn, big data stores many valuable and useful benefits and knowledge.
- adopting the idea of big data analysis as a recent trend in crisis response.
- the need to integrate the marketing information system with other information systems in the Algerian Telecom Corporation in order to avoid marketing crises.

## KEYWORDS

- **big data**
- **marketing crisis**
- **marketing information system**
- **telecom Algeria**

# REFERENCES

1. Feinleip, D., (2014). Big data boot camp. *Springer Science and Business Media Journal.* New York.
2. Al-Aklabi, A. A., (2018). Big data and decision-making. *Journal of Information and Technology Studies* (pp. 1–22). Bin Hamad Khalifa University Press.
3. Bin, T. Z., & Al-Rifai, S. B., (2018). New roles of information specialist to deal with big data. *Journal of Information and Technology Studies* (pp. 56–67). Bin Hamad University Press.
4. Al-Ajarmeh, T., & Al-Taie, M., (2002). *Marketing Information System* (p. 15). Al-Massira edition, Jordan.
5. Abu, Q. A. S., (2007). *Marketing, Modern University Office* (p. 129). Dar Al-Nahza Al-Arabi, Egypt.
6. Zaki, K. A., & Sayyed, I., (2008). *Marketing and Marketing Information* (p. 45). Zahran Publishing House; Jordan.
7. Abdel, M. M., & Al-Sherini, A., (1999). *Marketing Research and Marketing Information Systems* (p. 232). Dar Al-Nahza Al-Arabi; Egypt.
8. Guards, L., (2006–2007). *Running the Institution in Crisis* (p. 29). Master Thesis, University of Boumerdes, Algeria.
9. Joan, M. I., (2002). *The Marketing Crisis and How to Deal With it (The Role of Public Relations in Addressing It)* (p. 3). Master Thesis, Damascus University.
10. El-Hawary, S., (1999). *Marketing Crisis Management* (p. 83). Ain Shams Library; Part II; Egypt.
11. Ahmed, I. A., (2002). *Crisis Management Causes and Treatment* (p. 35). Dar Al-Fikr Al-Arabi, Egypt.
12. Abu-Juma, N. H., (2003). *Unhealthy Marketing Phenomena in the Arab World, the Second Arab Forum* (p. 47). Qatar.

# CHAPTER 17

# Perspectives of Big Data Analytics' Integration in the Business Strategy of Amazon, Inc.

MUSTAPHA BOUAKEL[1] and AMINA ZERBOUT[2]

[1]Associate Professor, Faculty of Economics, Commerce, and Management Sciences, University Center Ahmed Zabana, Relizane, Algeria, E-mail: mustapha.bouakel@univ-sba.dz

[2]PhD Student, Faculty of Economics, Commerce, and Management Sciences, University Ali Lounici, Blida 2, Algeria, E-mail: ea.zerbout@univ-blida2.dz

## ABSTRACT

This study aims to shed light on the role of big data (BD) in the promotion of Amazon's e-commerce. The study concluded that the success of the company is due to the early adoption of the principles of BD, and its integration into the commercial strategy.

The company has also been able to invest heavily in the development of applications for the big data mining (DM) and analytics, as well as designing a technical map that responds to the concerns of customers, gives greater ability to track their behaviors and protect the procedures of shopping, delivery, and payment, which indicates that Amazon's success in leading e-commerce involves standards of the control degree over the vast amount of data, the team skills and the nature of the technical tools to achieve its strategic vision. It is also expected that big data entry will continue breaking new horizons in this area in the foreseeable future.

## 17.1   INTRODUCTION

Big data analytics has emerged as a new interface for innovation and competition in a wide range of e-commerce. It provides a large number of companies that use people's dynamics, processes, and technologies to transform data into insights to make rational decisions and solutions to business problems.

This approach requires close integration of data, resources, skills, and systems in order to create a competitive advantage. Major e-commerce companies such as Google, Amazon, and Facebook have recognized big data analytics' role in driving their annual revenue increase.

Leading e-commerce company Amazon and its subsidiary Zappos were ranked among the top ten retailers at the American Express Awards Customer Choice Awards for 2010 and 2011, respectively [1].

Industry observers have acknowledged that recognition of the company's dominance in this area was the result of Amazon's use of big data resources to provide superior quality of service.

Since its emergence as the dominant Internet service provider in the early 2000s, Amazon began to focus on big data to improve its performance along with many major Internet companies. Since then, it has focused on properly utilizing huge databases about people who were shopping on e-commerce portals.

The acquisition of Zappos in 2009 facilitated the use of big data to improve customer service quality and verify organizational fraud. In addition to easy access to customer profiles, purchasing habits, and tracking of the browsing, it also facilitated the company's executives to provide quick solutions to customer concerns.

Numerous research has emerged trying to understand how Amazon's leading e-commerce company is using big data resources to improve its performance, especially in light of the overlap of many variables such as the breadth of its database of tastes and preferences, customers' previous purchase history, and divergent trends, in addition to the intense competition from other companies such as Ali Express and e-Bay and others.

Nevertheless, Amazon has succeeded in leading global e-commerce by analyzing big data and reflecting its vision to the extreme. On this basis, the problem statement can be raised in the following key question:

> *How did Big Data Analytics contribute to the success of Amazon's business strategy?*

The objective of the present work is to analyze how Amazon uses its own big data resources to build a better relationship with its customers, and appreciate the importance of developing big data capabilities to improve company performance. In addition to demonstrating Amazon's Strategy for helping other e-commerce portals to take advantage of big data resources, and explore ways in which Amazon can use big data resources to improve its business prospects.

Amazon's big data technologies have provided important opportunities to understand customers' preferences in detail and gain a competitive advantage in the market; it also allowed them to respond to the requirements of different segments and greater ability to track customer behaviors.

This has driven it to invest heavily in developing data management structures from the traditional data storage model to more complex structures such as real-time processing and batches, handling structured and unstructured data, and challenging high-speed transactions.

To understand the topic and answer the problem of the study, a descriptive approach was adopted in reviewing the basic concepts related to big data, its forms and methods of treatment, and the importance of exploiting them to give Amazon a competitive advantage, in addition to highlighting the role of technical tools in promoting its business strategy.

The study was structured into three main sections:

- Motives and milestones of transformation to deal with big data;
- Characteristics and backgrounds of Amazon's business strategy; and
- The role of big data in promoting Amazon's e-commerce.

## 17.2 MOTIVES AND MILESTONES OF TRANSFORMATION TO DEAL WITH BIG DATA

Big data represents the most important contemporary bets that have attracted the attention of managers and entrepreneurs, as a flexible and integrative approach to the various activities of the organization, especially opportunities to develop business models and employ the outputs of digital innovations that allow enhancing competitiveness and anticipating the future to ensure growth and continuity.

## 17.2.1   BIG DATA CONCEPT, FORMS, AND FEATURES

This term is often used in parallel with business intelligence (BI) and data mining (DM), but big data varies in size, number of transactions, number of sources, degree of dispersion, complexity, and accelerated generation. This requires special techniques in order to address and treat them [2].

Andrea defines big data as information assets characterized by high volume, speed, and diversity that create the need for advanced techniques and analytical methods to convert them into value [3].

The data in terms of their composition can be divided into the following [4]:

- **Structured Data:** Includes structured data, which is easily processed. This data is easily entered, stored, and analyzed. Structured data is stored in rows and columns and easily managed.
- **Unstructured Data:** These are disorganized and scattered data that flows from multiple and different sources and need to be organized, arranged, and structured in order to process and exploit them to obtain information useful in rationalizing decisions.

Big data refers to large and sparse dynamic data sizes created by people, tools, and machines that require new, innovative, and scalable technology to collect, store, and process the vast amount of data, which includes business movements, consumer behavior, risk, performance, product management and more.

Big data is made up of information from social media, data from devices that support the Internet (including smartphones and tablets), and data machines, video and audio recordings, and continuous storage [5].

Studies combine that big data is characterized by what is called 5V's [6, 7]:

1.  **Volume:** The amount of global digital data generated, replicated, and consumed in 2013 by the International Data Corporation (IDC) was estimated at 4.4 Zettabytes and doubling every two years.
    By 2015, digital data has grown to 8 ZB and is expected to grow to 12 ZB in 2016. 1 ZB is enough to store 200 billion high-definition movies that will take 24 million years to watch.
2.  **Variety:** In addition to numbers and text, there are data for image, audio, video, meteorological, and astronomical data, text, and e-mails

in blogs and social media, in addition to buying, selling, exchange, financial transactions, surveillance cameras, music, and more.

3. **Velocity:** It means that it is a process that runs overtime and never stops. These data are considered temporary and need to be analyzed when they are generated, with a view to exploiting them and creating value.

4. **Veracity:** Big data needs to be organized, structured, and reliability tested.

5. **Value:** The data analyzer is able to achieve high levels of value under big data, thanks to increased computing power, the advancement of programming languages, and increased storage capacity.

### 17.2.2   BIG DATA TOOLS

There are five main ways to analyze big data and generate added value, which can be summarized below [8]:

- **Exploration Tools:** Useful throughout the information lifecycle for rapid and intuitive observation of phenomena and analysis of information from any combination of structured and unstructured sources, these tools allow for analysis alongside traditional BI systems. Since there is no need for initial modeling, users can draw new insights, reach meaningful conclusions, and make informed decisions quickly.

- **Business Intelligence (BI) Tools:** It is important for reporting, analysis, and performance management, primarily with transaction data collected from data warehouses and production information systems. These tools provide comprehensive business evaluation and performance management capabilities, including enterprise reporting, dashboards, custom analysis, scorecard development, and analysis of various scenarios.

- **Tools within Databases:** It is a variety of techniques for finding relationships between certain variables, and given the application of these techniques directly within the database, it eliminates data traffic to and from other analytical servers, accelerating the information cycle and reducing overall cost.

- **Hadoop:** Useful for pre-processing data to identify partial trends or find interfaces, such as out-of-range values. It also enables companies to unlock the potential value of new data using servers.

- **Decision Management:** It includes predictive modeling, business rules, and self-learning for informed action based on the current context. This type of analysis allows individual recommendations to be valued across multiple channels, increasing the value of each customer's interaction.

### 17.2.3   BIG DATA TOOLS PROCESSING

The processing of big data by sequence is based on two main inputs:

1. **Data Mining (DM):** It is an inductive, iterative, and interactive discovery search in large databases of valid, new, useful, and understandable data models [9]:

   - **Iterative:** Requires several passes.
   - **Interactive:** User in the process loop.
   - **Fit:** Applicable now and in the future.
   - **New:** Unpredictable.
   - **Handy:** Allow users to make decisions.
   - **Concept:** Simple presentation.

   This process automatically extracts predictive information from large databases and understands the rules, relationships, and dependencies between variables through statistical and mathematical methods and recognition patterns [10].

2. **Data Analytics:** The science of examining raw data to draw conclusions about that information. Data analytics are used in many industries to allow companies and organizations to make better business decisions and in science to validate or deny existing models or theories.

   In any big data setup, the first step is to capture a lot of digital information about analytical excellence that leads to better decisions. The majority of raw data, especially big data, it doesn't offer much value in its raw condition. Thus, by applying the right toolkit, we can draw strong insights from this inventory of data [11].

## 17.3  CHARACTERISTICS AND BACKGROUNDS OF AMAZON'S BUSINESS STRATEGY

Amazon is the leading e-commerce leader and has grown significantly since its inception as an online bookstore as it sought to penetrate other areas such as logistics, consumer technology, and cloud computing. This has made its strategy more successful and differentiated in the field of electronic commerce and experience worthy of study, especially the optimization of design controls and the degree of adoption of BI and the exploitation of digital innovation outputs.

### 17.3.1  THE EMERGENCE OF AMAZON

Amazon was founded in 1994 by Jeffrey Preston Bezos and was dubbed Cadabra. It started operations at a time when the Internet was growing, and the Internet was considered a potential business medium.

By understanding this trend, Bezos came up with the idea of selling books online. He felt that books were the best products sold online, where millions of titles were printed. An e-commerce site can incorporate and sell more books than traditional libraries.

According to Bezos, the latter cannot have more than 200,000 books at a time and aim to build a huge online library that will be larger than any physical library in the world.

Amazon was initially funded with money borrowed by Bezos from friends and relatives. Bezos and his wife, along with some of the staff, build and test your website for more than a year before launching it. Amazon was finally opened to customers in 1995.

Like many other technology giants, it was initially operated by the Bellevue garage in Washington and expanded with the help of Dot-Com. At the time Amazon started operations, the retail market was very fragmented, and there was no major player except Barnes and Noble, one-tenth of the company's total market share but does not exist online.

Amazon got the first engine advantage and faced very little competition in the early days of its operation. Since starting its business, Bezos has focused on customers and believe that customer loyalty is the key to penetrating the market and increasing sales. Amazon began shipping goods to all 50 states in the USA and 45 other foreign countries within a month of launch, and all this was still working in the Bezos garage.

Amazon's popularity has grown through an oral talk where customers have recommended it to others. Within four months of its launch, Amazon sold more than 100 books a day [12].

Therefore, Amazon is a multinational e-commerce company located in Seattle, North America, and is a web giant, grouped under the acronym GAFAM along with Google, Apple, Facebook, and Microsoft. The company bears the slogan 'Work Hard, Have Fun, Make History' [13].

Amazon.com is the world's largest online retailer, trading in furniture, clothing, electronics, music, and many other commodities, Amazon's original specialty was selling books, but it diversified into selling other products, first by expanding cultural products available for purchase, and then through the gradual availability of products of all kinds.

Currently, even food products can be ordered via Amazon, and the company was listed on the Nasdaq Stock Exchange in May 1997 [14].

In 2018, the company had a market capitalization of $1,000 billion, a turnover of $177.9 billion (2017), its net profit exceeded $2.37 billion (2016), and the company employs 631,300 people worldwide (2018). In addition to the original US site approved in 1995, new sites are available in many countries.

### 17.3.2   EVOLUTION OF AMAZON'S ECONOMIC INDICATORS

The time series shows net revenue from e-commerce sales and Amazon services from 2004 to 2018, in billions of US dollars. Last year, the company's net revenue was $232.89 billion, up from $177.87 billion in 2017.

As of 2017, the company generates the majority of its net revenue through online retail product sales, followed by third-party retailer services, and retail subscription services, including Amazon Prime and AWS. Amazon is also the leading eBook reader product Amazon Kindle (Figure 17.1).

Amazon has employed 647,500 full-time and part-time employees. The list of products and services sold on Amazon includes a variety of electronics, books, cloud infrastructure supplies, clothing, furniture, food, toys, jewelry, and more. About 61% of Amazon's global revenue for 2017 was collected from electronic retail sales of electronics such as smartphones, tablets, and others, and about 17.9% came from third-party retail services.

Amazon has far more Google employees and competitors in e-commerce than eBay. According to the ranking of the most valuable companies in the world in 2018, Google ranked first, followed by Apple in second place with a brand value of \$300.59 billion, followed by the strongest companies such as Netflix, Publix, and Amazon.

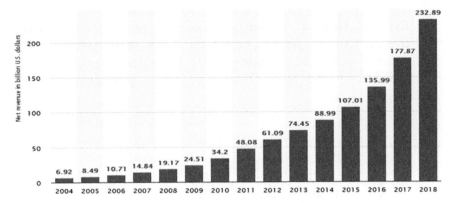

**FIGURE 17.1**   Net e-commerce sales revenue for Amazon 2004–2018.
*Source:* www.statista.com/topics/846/amazon.

On the other hand, Amazon's marketing spending on online retailers amounted to more than \$13 billion, an increase of more than \$10 billion compared to 2017. In 2019, the online retail platform was expected to account for 13.7% of e-retail sales worldwide. In the US, Amazon is expected to account for more than half of the domestic e-commerce market.

According to the survey conducted in 2017 on the characteristics of the company Amazon, 66% of respondents agreed that the company is sensitive to modernity, 62% of them admitted to being one of the most attractive companies for customers, while 55% of them saw it as an innovative company, while 24% saw it is as an environmentally friendly and adheres to the principles of transparency [15].

### 17.3.3   AMAZON'S BUSINESS STRATEGY

Amazon has applied the traditional retail recipe: wide choice, irreplaceable user experience, and low price. However, the genius of J. Bezos prompted him to apply this recipe to e-commerce by taking advantage of the benefits of the digital environment.

Unlike the traditional retailer Walmart or Carrefour, Amazon's business model offers certain advantages that are not allowed for offline trading, which has earned it valuable competitive advantages.

Amazon has resorted to several strategies to expand its business across the world and win many customers and suppliers, focusing on three levels [16]:

- **External Growth:** 16 categories of new products and services created. Example product categories: books, motorcycles, jewelry watches, sports, and entertainment, etc. Examples of new services are Cloud Computing, Market Place, and Amazon Web Services (AWS) (Figure 17.2).

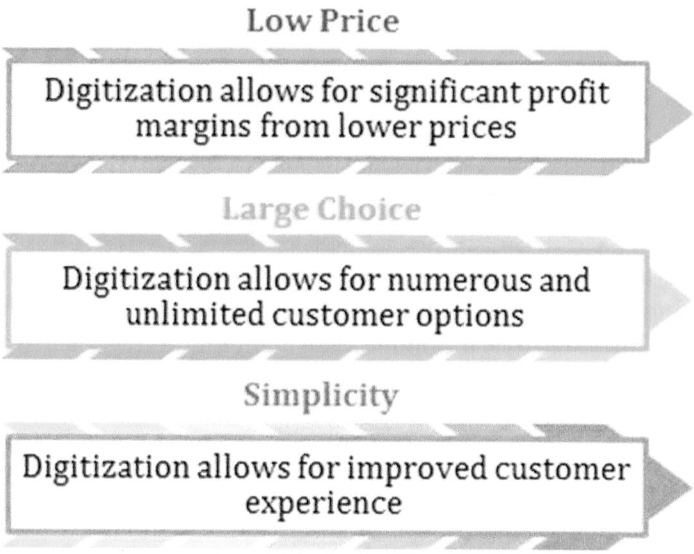

## Low Price

Digitization allows for significant profit margins from lower prices

## Large Choice

Digitization allows for numerous and unlimited customer options

## Simplicity

Digitization allows for improved customer experience

**FIGURE 17.2**    Amazon's business strategy.
*Source:* Roumieu et al. [16].

- **Acquisition of Competitors based in Strategic Markets, where**:
  - **1998:** Acquisition of the famous German bookstore Telebook and turning to Amazon.de, in addition to the English site Book pages and turning to Amazon.co.uk.
  - **1999:** Acquisition of PlanetHall.com to manage address records and customer phone numbers.
  - **1999:** 54% of the shares of the e-commerce site Pets.com, which specializes in the distribution of animal products (food, supplies, and equipment, etc.), purchased.

- o **1999:** Acquisition of Leep Technology Inc. specializing in the programming and development of customer relationship management systems.
- o **2004:** The acquisition of the largest Chinese e-commerce site Joyo.com specializing in the distribution of books and CDs, worth $75 million.
- o **2010:** The acquisition of QuidSi for $545 million, and allocated its first division Diapers.com to distribute diapers and the second division Soap.com for hygiene products.
- • **Alliance and Investment:** Amazon tried to control the horizontal and vertical markets through:
  - o **1999:** Alliance with Drugstore.com specialized in health products and cosmetics.
  - o **1999:** Merger with Accept.com, specialized in the development of B2B and C2C exchanges.
  - o **2000:** Purchase of stakes from Basis Tech, specialized in improving customer experience.
  - o **2000:** $60 million investment in Kozmo.com, an online delivery specialist.
  - o **2002:** A partnership agreement to develop the CDNow platform.

Adapa and Debapratim also added that transformation strategies included the following [17]:

- • **July 2002:** Amazon launches web services.
- • **June 2003:** First service branch established.
- • **April 2004:** The first jewelry store.
- • **February 2005:** Amazon Prime launches.
- • **September 2006:** music and video upload service.
- • **November 2007:** Amazon Kindle, November 2008, Free Packaging Initiative.
- • **May 2009:** Kindle DX service.
- • **April 2010:** The introduction of a new HQ strategy.
- • **July 2011:** The market value of $100 million.
- • **February 2012:** Launch of a sports store.
- • **August 2013:** Buy Washington Post.

## 17.4   ROLE OF BIG DATA IN PROMOTING E-COMMERCE FOR AMAZON

The widening of the customer base is the biggest challenge facing leading companies, but Amazon's integration of big data analytics into its business strategy has helped to set trends and better adapt to contemporary developments. It also left a more positive impact on the level of financial revenue and reduced costs but extended to create a strong competitive advantage and a broader vision about its future aspirations.

### 17.4.1   AMAZON DATA INCREASING FACTORS

In 2016, Amazon recorded more than 304 million active client accounts, 65% of which were female, and received an average of 4,310 visits per minute. Amazon Prime members spend an estimated $1,500 on average, while other customers spend about $625 annually; in addition, members have access to over 1 million e-books.

Statistics show that the company has reduced prices for year-end offers (2015) for 489.5 million products, and acknowledged that 70% of purchases made by smartphones. Amazon's market share during Black Friday's was around 35.7%, while shipping costs were estimated at $1.54 billion, which makes the company actually seeking to gain the best by looking for mechanisms to integrate and exploit the entrance of big data analytics to push and develop its strategy and make a qualitative leap in the field of electronic commerce [18].

### 17.4.2   AMAZON'S BIG DATA TOOLS

AWS provides a wide range of managed services to quickly and easily create comprehensive, secure data applications, whether applications require physical flow or aggregated data processing.

AWS provides the infrastructure and tools to process a big data project by collecting, storing, and analyzing it. AWS has an ecosystem of analytical solutions specifically designed to handle this growing amount of data and provide a comprehensive view of the company's business model.

Amazon's white paper lists the technical tools used to handle big data [19]:

- **Amazon Kinesis:** It is an AWS data-streaming platform, making it easy to download and analyze streaming data, and provides the ability to create custom data streaming applications for specialized needs. Using Kinesis, you can accommodate real-time data such as application logs, site clicks, IoT telemetry data, and more in databases, data lakes, and data warehouses, or create real-time applications using this data. Using streaming data allows real-time dashboards to be triggered, alerts generated, pricing, and advertising implemented Dynamic. Amazon Kinesis enables data processing and analysis as it arrives and responds in real-time, rather than having to wait until all data is collected before processing can begin.
- **AWS Lambda:** It allows code to run without providing or managing servers, the company pays only to calculate the time it consumes, and there is no charge when the code is not running. With Lambda, you can run code for any type of application or backend service-all without management. The company engineer raises the company code while Lambda handles everything needed to run it and expand the code. The code can also be set up to run automatically from other AWS services or connect to it directly from any web or mobile app.
- **Amazon EMR:** It is a highly distributed computing framework for processing and storing data quickly, efficiently, and at a low cost. Amazon EMR Apache Hadoop, an open-source framework, is used to distribute and manipulate data across a resizable set of Amazon EC2 instances and allows you to use the most common Hadoop tools such as Hive, Pig, Spark, etc. Hadoop provides a framework for running massive data processing and analytics. Amazon EMR does all the work involved in providing, managing, and maintaining Hadoop Group's infrastructure and software.
- **AWS Glue:** It is a fully managed ETL service that can be used to reliably index, clean, enrich, and transfer data between data stores. With AWS Glue, the cost, complexity, and time it takes to create ETL functions can be greatly reduced.

In another white paper for 2019, Amazon added [20]:

- **Amazon Machine Learning:** It is a service that makes it easy for anyone to use predictive analytics and machine learning technology. Amazon ML provides browsing tools to guide you by creating machine-learning models (ML) without learning complex ML algorithms and techniques. The embedded processors are guided through interactive data exploration steps and trained in the ML model to assess the model quality and adjust outputs to match business objectives. After the model is ready, predictions can be requested either in batches or by using a low real-time API.

- **Amazon Athena:** It is an interactive query service that makes it easy to analyze data on Amazon S3 using standard SQL. There is no installation or management infrastructure. You can start analyzing data immediately, and you don't need to upload your data to Athens, as it works directly with the stored data In S3. It only requires logging into the Athena console, defining the table schema, and starting the query.

- **Amazon DynamoDB:** It is a fast, fully managed NoSQL database service, which makes storing and retrieving any amount of data simple and cost-effective, and serves any level of application traffic. DynamoDB helps to unload the administrative burden of operation and expand the database in a more distributed and accessible manner.

- **Amazon Redshift:** It is a fast, large, and fully managed data warehouse storage service that allows you to analyze all data efficiently, effectively, and at low cost using BI tools. Optimized for data collection ranging from hundreds of gigabytes to beta bytes or more, designed to cost less than one-tenth of the cost of traditional data storage, this system automates most common administrative tasks associated with supplying, configuring, monitoring, backing up, and securing the data warehouse, making it easy to manage and inexpensive. This automation allows the construction of petabytes-sized data warehouses in minutes, rather than weeks or months taken by traditional local implementations.

- **Amazon Elastic Search:** The flexible search makes it easy to deploy, operate, and extend searching for log analytics, full-text scanning, application monitoring, and more. AES is a fully managed service that provides easy-to-use APIs for flexible search and enhances availability, expansion, and security according to the requirements of the production environment. The service offers implicit integrations with previous services to quickly move from raw data to actionable insights.

- **Amazon Quick Sight:** It is a fast, easy-to-use business analytics service running on the cloud, making it easy for all employees within an organization to create visualizations, conduct custom analysis, and quickly get business insights from their data anytime, on any device.

It can connect to a wide range of data sources, including flat files, for example, CSV and Excel to access logical databases, including SQL Server, MySQL, and PostgreSQL.

Organizations extend their business analytics capabilities to hundreds of thousands of users and provide fast, responsive query performance using a powerful in-memory engine (SPICE).

Ian et al. also added [21]:

- **Amazon EC2:** It provides an ideal platform to run large self-managed data analysis applications on AWS infrastructure. Almost any software can run on Linux or Windows virtual environments on Amazon EC2, and you can use the Pay As You Go pricing model, and there are many options for self-managed big data Analytics.

### 17.4.3   VALUATION OF DATA MINING (DM) AND BIG DATA ANALYTICS

With millions of customers worldwide, Amazon is the largest online store offering amazing products and services online. They have a great customer database and use this data to build strong relationships with their customers.

By analyzing and summarizing useful information about customers, they design their strategies for promotions and product improvements, as well as incorporating the concept of extracting data from the supply chain into marketing processes [22].

Amazon also uses DM to market its products in various aspects to gain a competitive advantage. Customers want customization from companies they buy products, most online companies, because of increased social media interventions [23].

Providing targeted information to a customer service representative who deals with a particular customer is a required opportunity for development. Amazon Customer Representatives have access to complete customer data and can analyze and discuss the problem once a call is received from the customer, so customers are satisfied that they will be taken into account with their orientation.

Smart retailers such as Amazon are also characterized by the efficient use of data collected through efficient sources, and the results are exploited in a way that creates value for the company. Customers can also control the information they want to share or not, giving them a sense of ownership and control.

Customer data has become a way to build strong brands and customer loyalty through effective DM, where each client is treated individually and prioritized in building the company's strategy. They feel comfortable interacting with them if they get control and transparency.

Marketers get information with every click. Amazon's customer purchase history helps them determine customer preferences and options [24]. Then they make the advertising campaign according to the customer's choice. Social data traffic is much faster than manual data and is automatically downloaded. Behavior patterns are studied in identifying marketing channels and developing marketing strategies.

Amazon recently patented a system designed to ship goods to customers before deciding to buy them (predictive sending); it is a strong indication of their increased confidence in reliable predictive analytics.

An important factor to consider when looking at Amazon is the extent to which its big data is traded, compared to that of other companies that deal with data on a similar scale. Unlike Facebook, which may know a lot about movies you like or about your friends, the vast majority of Amazon's customer data relates to how they attract hard cash [1].

With 360° technology developed using massive data resources to create customized marketing messages for customers, customers have been able to learn more about products. Apart from improving its business performance, big data resources have also been used for some innovative uses, such as preventing theft of goods from its warehouses. Amazon also allowed smaller e-commerce companies with limited resources to use big data services.

In 2010, Amazon launched an innovative service called Amazon Webstore, which allows smaller businesses to build shopping portals based on Amazon's e-commerce platform [25].

## 17.5  CONCLUSION

Amazon has evolved beyond its original beginning as an online library, thanks to its early adoption of big data principles and its integration into its business strategy. It has also invested heavily in the development of exploration applications and analysis of big data.

In addition, the company has been able to design a technical map that responds to the concerns of customers and gives greater ability to track their behavior and protect the procedures of shopping, delivery, and payment, etc., which indicates that Amazon's success in leading e-commerce involves related to the degree of control over the vast amount of data, the level of teamwork skills and the nature of the technical tools harnessed to reflect its strategic vision.

The study concluded that Amazon's business strategy is based on targeting low-profit margins as a result of price reductions, diversification of options, and the development of mechanisms to improve the customer experience. In addition, Amazon's technical tools are characterized by diversity, flexibility, and integration of applications; this has resulted in excellent data management control and more comprehensive and effective analysis of their customer data.

The integration of big data into Amazon's business strategy has contributed to generating high added value by creating more polarized responsive outputs for customer concerns. In its current move, Amazon aims to consolidate and make available its cumulative experience by selling aggregation services, processing, exploration, and big data Analytics for startups.

### 17.5.1   LIMITATIONS OF THE STUDY

Big data is formed through the process of collecting data of people who have already visited or revert to Amazon's website periodically, so the database does not in any way reflect the general trend of customer behavior and still need new mechanisms to collect and monitor the mobility of the people out of contact.

Exploration of big data can also produce misleading and non-significant (expressive) results, since in some cases the data can be generated from a single source but for multiple customers, in addition to many sources of confusion that occur extreme and anomalous data, which calls for more efficient methods of diagnosing data before analyzing it.

The outputs of big data Analytics are relatively significant and inclusive. Thus, it drives Amazon to the need to correct and check the degree of uncertainty in a way that minimizes the risk of assuming that all outputs are absolutely applicable to the user base.

As much as the vast amount of data gives Amazon a broader vision, it is also a major impediment to its graphical representation and in a way that allows it to understand the background of the creation of such data, the motives for its generation, and ultimately to make more rational decisions.

Amazon's targeting of a very wide market base makes data difficult to classify and organize, especially trying to elucidate its dimensions since it deals with categories that vary in terms of tastes, living standards, religions, and others. This requires to take into account other variables related to the volume of expenditure, degree of openness, quality of products, and the general level of prices in each region.

The collection of big data is linked to the extent to which customers are authorized to track their behavior through the digital environment. Therefore, the future of investment in this field depends on the company's ability to secure users' data and convince them to access and monitor their mobility in all circumstances, all the time, and in any space.

Technical tools, such as machine learning software, are sometimes irrelevant to smaller data, although they have a strong and important business significance that reflects the orientation of a minority of customers, and can make a difference and an additional shift in Amazon's strategy if they are taken into account.

Exploring and analyzing big data is time-consuming and costly. In contrast, user behaviors are characterized by acceleration and dynamism, which raises the problem of synchronizing Amazon's adaptation of its policies in real-time to customers' trends and influencing their purchasing decisions.

## 17.5.2  FUTURE RESEARCH

Exposure to the present study has given rise to problems closely related to the above limitations, which can open up multiple perspectives to address the mechanisms and methods of bridging the gap.

On the other hand, the current study believes that the optimal field for future research could target how to integrate the human element in big data processes since the technology and software alone are not able to process information securely and allow to reach more meaningful and rational decisions, as well as the human element's supervision of the mining and analysis processes helps to straighten the flow of data and understand the backgrounds, motives, and perspectives of their creation.

## KEYWORDS

- **Amazon**
- **big data**
- **commercial strategy**
- **data analytic**
- **data mining**

## REFERENCES

1. Marr, B., (2015). *Big Data: Case Study Collection, 7 Amazing Companies That Really Get Big Data* (p. 29). Wiley, New York.
2. Xiaomeng, S., (2012). *Introduction to Big Data* (p. 11). Knowledge for a better world, Department for informatics and e-learning at NTNU.
3. DeMauro, A., Greco, M., & Grimaldi, M., (2014). What is big data? A Consensual definition and a review of key research topics. *International Conference on Integrated Information (IC-ININFO), AIP Proceedings, 97–104.*
4. Shagufta, P., & Umesh, C., (2017). Influence of structured, semi-structured, unstructured data on various data models. *International Journal of Scientific and Engineering Research, 8*(12), 67–69.
5. Sharma, R., (2016). Big data and analytics in the audit process. *Board Matters Quarterly* (Vol. 9, p. 20). Ernst & Young LLP, Germany.
6. Ishwarappa, & Anuradha, J., (2015). A brief introduction on big data 5Vs characteristics and Hadoop technology. *International Conference on Intelligent Computing, Communication and Convergence (ICCC-2015), Procedia Computer Science, 48,* 319–324.
7. Rajaraman, V., (2016). *Big Data Analytics* (Vol. 21, No. 8, pp. 695–716). General Article, Springer.
8. An Oracle White Paper, (2013). *Big Data Analytics: Advanced Analytics in Oracle Database* (p. 13). Oracle Corporation; USA.
9. Talbi, E. G. *Data Mining: A Tour of the Horizon* (p. 325). Lille Fundamental Computer Laboratory.
10. Gilles, G., (2016). *Introduction au Data-Mining* (p. 30). INSA Rouen-Département ASI Laboratoire LITIS.
11. Sofiya, M., & Aishwarya, J., (2015). Data analytics types, tools and their comparison. *International Journal of Advanced Research in Computer and Communication Engineering, 4*(2), 488–491.
12. Amazon's Big Data Strategy, (2014). *ICRM-IBS Center for Management Research* (p. 15). Hyderabad, India.
13. Maylis, B. *Histoires De Géants (3/4): Amazon, UN Empire Logistique.* Disponible: https://www.franceculture.fr/emissions/entendez-vous-leco/trois-histoires-de-geants-economiques-34-amazon-livraison-a-la-chaine, (accessed on 22 October 2020).

14. Faherty, E., Huang, K., & Land, R., (2017). *The Amazon Monopoly: Is Amazon's Private Label Business the Tipping Point?* (p. 29). Munich Personal RePEc Archive, Bentley University PMBA Program, McCallum Graduate School of Business.
15. The Statistics Portal, (2020). www.statista.com/topics/846/amazon (accessed on 22 October 2020).
16. Roumieu, Kahit, Aubry, & Baivier, D. F., (2014). *Etude du Business Model Amazon* (p. 73). E&A Etudes & Analyses, Panthéon-Sorbonne, Université Paris 1.
17. Adapa, S. R., & Debapratim, P., (2003–2019). *Amazon's Big Data Strategy* (pp. 67–77) Chegg Study. Retrieve from: https://www.chegg.com/homework-help/questions-and-answers/amazons-big-data-strategy-case-analysis-1-executive-summary-2-swot-analysis-3-problem-stat-q18882320 (accessed on 22 October 2020).
18. *Amazon Facts and Figures*, (2016). https://whatsthebigdata.com/2016/04/05/amazon-facts-and-figures/ (accessed on 22 October 2020).
19. AWS White Paper, (2018). *Amazon Web Services-Big Data Analytics Options on AWS* (p. 56). AWS White Paper.
20. AWS White Paper, (2019). *Building Big Data Storage Solutions (Data Lakes) for Maximum Flexibility* (p. 29). AWS Whitepaper.
21. Meyers, I., Elliott, D., & Batalov, D., (2015). *Large Scale Data Analytics on AWS* (p. 111). Amazon Web Services, PPT Projection.
22. Dholakia, R. & Dholakia, N., (2013). Scholarly research in marketing: Trends and challenges in the era of big data. *International Encyclopedia of Digital Communication and Society*, 2–31, Retrieved from: http://web.uri.edu/business/files/Encycl-Communication-DataMining-n-Marketing-.pdf (accessed on 22 October 2020).
23. Thabit, Z., (2015). Data mining by Amazon. *International Journal of Scientific and Engineering Research, 6*(6), 867–868.
24. Cinfia, J., Bunzel, D., Scruggs, S., (2014). *Consumer Marketers: Digging Too Deep With Data Mining?* Digital wave. Retrieved from: http://digitalmediaix.com/consumer-marketers-digging-too-deep-with-data-mining/#.VXMgI8-qqko (accessed on 22 October 2020).
25. Chen, H.M, Schütz, R., Kazman, R., & Matthes, F., (2016). Amazon in the air: Innovating with big data at Lufthansa. *49th Hawaii International Conference on System Sciences, IEEE Computer Society*, 5096–5105.

# CHAPTER 18

# The Hospital Information System: A Fundamental Lever for Performance in Hospitals

ZINEB MATENE[1] and KHALIDA MOHAMMED BELKEBIR[2]

*[1]Assistant Professor and Researcher, Faculty of Business Economics and Management, Djillali Bounaama University, Theniet El Had Street, Khemis Miliana, W. Ain-Defla, Algeria, E-mail: z.matene@univ-dbkm.dz*

*[2]HDR, Senior Lecturer, and Researcher, Faculty of Business Economics and Management, Djillali Bounaama University, Theniet El Had Street, Khemis Miliana, W. Ain-Defla, Algeria, E-mail: k.mohammed-belkebir@univ-dbkm.dz*

## ABSTRACT

Hospitals have found it useful to have a hospital information system (HIS) to achieve the desired performance. Indeed, HIS is a tool that allows the automation of treatments, the preservation of data, the exchange of information, and the fast execution of tasks. The goal of this chapter is to highlight the role played by the HIS on hospital performance. We concluded that the HIS is one of the fundamental operators of hospital performance, through several benefits such as time-savings, reduction of errors, easy access to knowledge, productivity gains, and improved quality of care.

## 18.1 INTRODUCTION

Computer Science continues to invade the various fields of human activities. This is due to its indisputable contribution to those who use it. Indeed, this tool allows, among other things, process automation, and real-time information exchange.

For several years, care processes have become increasingly complex. This led to a subdivision of medical specialties, with the effect of increasing the costs of patient care, the impoverishment of the doctor-patient relationship, the lack of mastery of therapeutic processes, and an absolute need to find a better way to streamline medical and administrative information for health professionals.

In this context, information and communication technologies (ICTs) seem to be a vector promoting the coordination of the health professional, the optimization of health expenditure through a good organization of care processes, and cooperation to enable better patient care.

Thus, the hospital information system (HIS) is presented as a major tool for automation and information processing. Indeed, it is an integrated hospital information processing system that is necessary for the hospital's daily operation, its management, its evaluation, and planning.

This chapter aims to shed light on the role of the HIS in achieving hospital performance. Our goal is to try to provide some answers to the following question: What is the contribution of a HIS to hospital performance?

To answer this question, we divide our work into three main parts as follows:

> ➤ The first part presents a brief state of the hospital information system;
> ➤ The second part is devoted to the presentation of hospital performance; and
> ➤ The third part examines the contribution of a hospital information system to hospital performance.

## 18.2   HOSPITAL INFORMATION SYSTEM (HIS)

Before addressing the HIS, it is necessary to define the following concepts: system and information system.

### 18.2.1   DEFINITION

1. **System and Information System:** The system and the information system can be defined as follows [1]:

   • A system is defined as the realization of a correspondence between a set of input variables and a set of output variables.

- An information system is defined as that enables the acquisition, storage, processing, and communication of information circulating in establishments.

2. **Information System of a Hospital:** Information systems are now seen as an activity that contributes to the value created by an organization in general and the hospital in particular.

A HIS is an information system applied to health facilities (hospitals, clinics, etc.). It manages all the administrative and medical information of the hospital [1].

The HIS is inserted in the "hospital" organization in perpetual evolution. According to predefined rules and operating procedures, it is able to acquire data, evaluate them, process them by computer or organizational tools, and distribute information containing a strong added value to all internal and external partners of the establishment, collaborating in a common work directed towards a specific goal, namely the care of a patient and the recovery of it [2].

According to the definitions, we find that the HIS is a system that tracks, collects, analyzes, and processes information from hospitals to achieve their objectives.

## 18.2.2   OBJECTIVES

A hospital information system allows to [3]:

- Provide patients with the memory of their medical history and access to their data;
- Improve the care of patients by caregivers;
- Optimize the resources of the institution;
- Guarantee quality by improving decision-making, reducing, and preventing incidents;
- Improve the understanding of the problems related to human pathologies and the health system for research and epidemiological purpose.

The information system plays a significant role in achieving the objectives of hospitals. Indeed, the HIS contributes to [1]:

- A reduction of waiting times;
- Availability of information;

- Help with decision-making;
- A faster exchange between the different actors (doctor-doctor, doctor-medico-technical staff, doctor-nurse, etc.).

Indeed, it allows the patients' efficient administrative management, the registrations of the patients to the external consultations, the hospitalizations, the appointments, etc. A hospital can launch a computerization project, giving the means to achieve the following objectives [3]:

- Acquire high visibility on the activity of the establishment;
- Manage activities in a timely manner;
- Achieve a better distribution of medical information that becomes more fluid;
- Align the establishment with good hospital management practices;
- Reduce IT operating costs.

### 18.2.3  COMPONENTS

A hospital information system consists of the following main parts [4]:

1. **Medical System:** It includes the following functions: medical procedures, pathology, morbidity, epidemiology, etc.
2. **Medical-Administrative System:** It contains the following functions:

   - **Patient Management:** It includes admissions management, outpatient care, billing, medical imaging, and follow-up of laboratory, and biochemistry exams.
   - **Financial and Accounting Management:** It concerns the accounting management system, financial management, and the accounting of fixed assets.
   - **Wealth Management:** It contains the management of supplies and store stocks, the management of drug stocks, the monitoring and maintenance of the biomedical park, and the management of the stock of medical devices.

3. **Medical-Technical System:** It covers the following functions: pharmacy, laboratory, and medical imaging.

## 18.2.4   FUNCTIONS

The HIS must be designed to facilitate the real-time integration of information between the operational and the decision-making. The HIS can regroup several necessary functions such as [5]:

- Management of schedules;
- Management of the pay;
- Billing;
- Budget monitoring;
- Statement of medical activities;
- Communication (internet, intranet, protocols, messaging, forum, order form, etc.).

The current trend is towards the outside of the hospital, i.e., the development of health networks, personal medical records, telemedicine, and why not, the surgical control of a remote robot. For example, a HIS allows the management of the healthcare record and the drug circuit through the following functions [3]:

- Computerized prescription (drugs, specialized consultation, etc.);
- Nursing plan;
- Management and monitoring of targeted transmissions;
- Follow-up of medical observations;
- Validation of care and administration of drugs by staff;
- Coding of the acts and diagnoses of the patient;
- Complete management of the pharmacy, stocks, orders, and the therapeutic booklet;
- Complete traceability of medicines and blood derivatives.

## 18.3   Hospital Performance

Before addressing the performance of a hospital, it is significant to define the concept of Organizational Performance first.

## 18.3.1   DEFINITION

1. **Organizational Performance:** Performance is about fulfilling the mission, adapting to its environment, and benefiting from what it can

offer, and producing quality results that meet the needs of customers and the objectives of the organization.

It is also supposed to achieve the expected results at the best cost while respecting the organizational conditions related to the optimal use of resources and the quality of the process, for the performance to be efficient [6].

According to this definition, performance encompasses the following dimensions:

- Customers and their needs;
- Processes;
- Resources;
- Impacts;
- Environment.

2. **Performance of a Hospital:** According to the World Health Organization (WHO), a high-performance health organization should rely on professional skills that match current knowledge and available resources and technologies, the efficient use of resources, the minimization of risk for health the patient, patient satisfaction and health outcomes [7].

The concept of performance in the hospital environment is as follows [8]:

- A hospital facility is efficient if it achieves its objectives. In other words, the improvement of the health status of people going there, the accessibility to care and overall care, the quality of care and practices, the relevance of such practices, and the adequacy of the use of sources;
- Performance relates to the optimal use of available resources;
- Performance affects the hospital's ability to adapt and innovate in order to respond to technical and scientific developments.

### 18.3.2 OBJECTIVES AND COMPONENTS

The search for performance in a hospital is centered on three objectives [7]:

- Providing effective patient management processes;
- Optimizing activities and ensuring rational use of available resources (human, technical, financial); and
- Ensuring the availability of competent human resources.

Hospital performance is made up of several dimensions, such as effectiveness, equity, relevance, safety, accessibility, innovation, quality, and partner satisfaction [9]:

1. **Effectiveness Dimension:** In the health field, there is a distinction between medical efficiency and economic efficiency. The first assumes that the provision of care is consistent with the recommendations of qualified professionals, while the second focuses on the optimization of production given the resources allocated.

2. **Equity Dimension:** Equity is a concept that means the spirit of justice. Equity refers to the ability of the hospital to provide individuals and populations with comprehensive and high-quality services in an equitable manner.

3. **Relevance Dimension:** The relevance of hospital services is based on scientific knowledge to determine to whom the services can benefit the most. This favors a reduction of the incidence, the duration of the intensity, and the consequences of the health problems. Relevance refers to the adequacy of the practices delivered to the patients and the quality of realization of the different acts.

4. **Efficiency Dimension:** Efficiency refers to efficiency at the lowest cost, i.e., the ratio between the result obtained and the resources used. The efficient delivery of hospital services aims, among others, to avoid the waste of equipment, supplies, ideas, and energy. It refers to the minimization of resources spent to obtain services.

5. **Security Dimension:** Security reduces the risks of unexpected or harmful results.

6. **Accessibility Dimension:** Accessibility to hospital services refers to the ease with which individuals can contact hospitals within a reasonable time and distance.

7. **Innovation Dimension:** Innovation refers to the implementation of a new idea generated internally or borrowed elsewhere, in a hospital setting. This new idea can affect a product, a device, a system, a process, a policy, program, or service.

8. **Learning Dimension:** The learning capacity in a hospital corresponds to the efficiency with which one arrives to acquire and transfer knowledge within the institution, but also to modify the behaviors in the function of knowledge and new perspectives.

9. **Reactivity Dimension:** The concept of time and sequences adapted to support.

10. **Continuity of Care Dimension:** Continuity of care is of great importance in the context of complex care involving several structures; it contributes to the quality and safety of services.
11. **Quality Dimension:** Quality of care is a *Sine qua non* of performance.
12. **The Dimension of the Partner's Satisfaction:** The dimension of satisfaction is summed up in the ability of the hospital to meet the needs of its patients in terms of care and its employees, particularly in terms of motivation, improvement of the work conditions, and permanent education.

## 18.4   RELATION BETWEEN INFORMATION SYSTEM AND HOSPITAL PERFORMANCE

### 18.4.1   CONTRIBUTION OF HOSPITAL INFORMATION SYSTEM (HIS) TO HOSPITAL PERFORMANCE

The hospital information system contributes to hospital performance as follows [10, 11]:

1. **Saving Time:** HIS allows the hospital to save time for quick and efficient care of the patient via a reduction or elimination of transcriptions, a reduction of the duration of the cycle of complementary examinations, a reduction in clerical tasks performed by medical and/or nursing staff, easier access to medical data and shorter stays.
2. **Reduction of Errors:** HIS allows the hospital to reduce and limit inappropriate medical prescriptions, incomplete prescriptions, and errors in the transcription of results.
3. **Easy Access to Knowledge:** HIS allows the hospital to reduce the variability of medical behaviors.
4. **Productivity Gains:** HIS allows the hospital establishment to control the costs, via a reduction of the stay duration, a reduction of administrative tasks, and a decrease in peak activity and resource optimization.
5. **Improving the Quality of Care:** HIS allows the hospital to improve the quality of care offered by it. This improvement is based on enhanced communications, reduced wait times, and patient records integration.

## 18.4.2 TOWARDS A HIGH-PERFORMANCE HOSPITAL INFORMATION SYSTEM (HIS)

To achieve a high-performance hospital information system, the following conditions are necessary [3]:

1. **A Development of Incompetent Human Resources:** The HIS does not only work with technologies and components, but it also requires skills and know-how to implement them.
2. **Deployment of Hospital Information Systems (HIS):** One of the essential conditions for the modernization of hospitals in the development of the HIS. It is on this condition that the sharing of information is an essential element for safe and quality care.
3. **Quality of Information:** The quality of care and the reduction of identification errors require an adequate, robust, safe, and well-controlled device. As a result, the quality of information used to identify a person and the procedures for control and identity management are at the heart of the quality of the information system. The concern is centered on the patient's secure management.
4. **Achievement of Hospital Efficiency:** To achieve this hospital efficiency through HIS, it is necessary to mobilize the necessary skills. All decision-makers must be aware of productivity gains in efficiency and organization, in quality, in human resources and engineering.
5. **Aiming for Medical and Social Excellence:** Any hospital establishment aims for medical excellence and social excellence. If an isolated person has to be treated, his reintegration will likely be more complicated, and therefore the person will stay longer at the hospital. Information is also working on the management of downstream and upstream: home hospitalizations, retirement homes, etc.
6. **Encourage Dialog within the Hospital Establishment:** To build economic and social performance, it is necessary to resume the dialog between health professionals. In other words, being able to reflect together on care, pathologies, but also to prevention logic by involving stakeholders in a better understanding of what is happening to them or what can happen to them tomorrow. This implies a new dimension of governance within the hospital, but also within different networks of health actors. They need to talk to each other, to have meeting places, methods, and relational qualities.

## 18.5    CONCLUSION, LIMITATION, AND FUTURE RESEARCH

The purpose of this work was to update the contribution of a HIS to hospital performance.

As mentioned earlier, the HIS is a computer system designed to facilitate the management of a hospital's medical and administrative information.

The objective of the HIS is to optimize the management of care activity by improving the management of information within the medical-technical area and to reinforce the coordination of medical tasks, administrative, and logistics performed within the hospital.

The HIS is geared towards improving patients' health status and optimizing their care. It aims to optimize processes to achieve the goals of the hospital.

As a consequence, the HIS displays several advantages such as time-saving, reduction of errors, easy access to knowledge, productivity gains, and improved quality of care.

The implementation of efficient HIS requires a real hospital information policy conducted by an information system and organization management. It must be long-term, evolutionary, and realistic since computer technology constantly evolves. It is necessary to reconcile opportunistic and long-term visions. Technical solutions are one thing, but the most important challenge is to set up real management.

It should be noted, however, that our study remains essentially conceptual. We are still in need of a full-fledged empirical study in order to reinforce the results obtained through this article. Therefore, the following research agenda may be proposed:

- The implementation of an empirical study on HIS and hospital performance.
- Additional research on the design of a HIS in Algerian hospitals.
- Further analysis of IT investment in the health sector.
- Studying the impact of informatics on the redaction of complications and mobility in a hospital.
- Assessing the development of the computer system and the quality of care.
- The study of smart hospitals.

## KEYWORDS

- **hospital**
- **hospital information system**
- **performance**

## REFERENCES

1. Gnoumo, M. Y., (2005). *Establishment of the Hospital Information System of the Sandoe CLINIC* (p. 53). Higher school of computer science (Cycle of Design Engineers in Computer Science), Polytechnic University of Bobo-Dioulasso Burkina Faso.
2. Ponçon, G., (2000). *The Management of the His: The End of the Technological Dictatorship*. Edition of the National School of Public Health.
3. Brizé, N., (2008). *Hospital Informatics: Public Health Challenges* (p. 23). Synthese: Health Informatics Meetings, Seminar.
4. Miled, M., (2004). *The Hospital Information System: Management Applications to the Computerized Medical File* (p. 47). Expert- TIC-DG CIMSP, SIT EXPO 2004-E- health Conference, Casablanca, Morocco.
5. Wikipedia, (2019). *Hospital Information System*. https://fr.wikipedia.org/wiki/Syst%C3%A8me_d%27information_hospitalier (accessed on 22 October 2020).
6. Voyer, P., (2009). *Management Dashboards and Performance Indicators* (2nd edn.). Quebec.
7. Besombes, B., Marcon, E., & Albert, F., (2007). *Evaluation of the Performance: Development of a Dashboard of Assistance to the Piloting of the Medico-Technical Plateau* (pp. 261–269). In *Gestions Hospitalières* review, University Hospital Center of Saint-Etienne.
8. De, C. S., De, S. C., & Guidet, B., (2008). *Indicators for Measuring Hospital Performance* (No. 216, pp. 585–593). In *Gestions hospitalières* review.
9. Canadian Institute for Health Information, (2013). Framework for Measuring the Performance of the Canadian Health System. Montreal
10. Degoulet, P., (2001). *Hospital Information Systems* (p. 45). Faculty of Medicine Broussais-Hôtel-Dieu.
11. Bagayoko, C. O., (2010). *Establishment of a Hospital Information System in Francophone Africa: Study and Validation of the Model in Mali* (p. 142). Doctoral thesis in Life Sciences and Health, Laboratory of Teaching and Research on the Treatment of Medical Information, Mediterranean University Aix-Marseille II, France.

# PART IV

# Big Data and Sustainable Development

# CHAPTER 19

# Big Data Analysis and Sustainable Development

DEHBIA EL DJOUZI

*Senior Lecturer and Researcher, Faculty of Business Economics and Management, Djillali Bounaama University, Theniet El Had Street, Khemis Miliana, Algeria, E-mail: raison81@yahoo.fr*

## ABSTRACT

Through this chapter, we will explore how big data analysis supports sustainable development efforts. In fact, big data projects can make a significant contribution to health care by improving and reducing health services costs, preventing disease and supporting innovation, research, and development in medicines, treatment, and improvement of public health.

Big data also helps educational institutions in the understanding of their beneficiaries, their requirements, and the reduction of the educational process costs. Big data projects can also contribute to fighting crime, ensuring security, coping with and predicting natural disasters, as well as supporting environmental conservation efforts.

At the economic level, analyze the impact that big data can have on economic policy-making and rationalization of resource exploitation.

However, big data projects are not without challenges such as privacy issues, lack of human capacity, expanding IT infrastructure to deal with the demands of large data sets, as well as risks of growing inequality and bias. There are gaps between those who own data and those who do not.

But big data can deliver on its promises to achieve sustainable development in the coming years if more efforts are made to improve affordable access to ICTs and knowledge for all people, promote opportunities for open learning and lifelong learning, build human capacity to exploit big data, ensure that specialized expertise is kept up with technological developments and promote North-South, South-South cooperation for technology transfer and skills.

## 19.1   INTRODUCTION

Increased technological development has allowed the analysis of the vast amounts of available information on the Internet or the so-called "big data," which led to the development of new technologies that can be used in many areas such as business, medicine, education, and security.

Three factors contributed to their emergence; first, Internet users exchanged an astronomic amount of data related to cross-cutting topics; second, the growing capacity of data centers to store increasing volumes; third, effective cloud computing can process trillions of digital processes just in few seconds.

Simultaneously, the scrutiny of "big data" has led to obtain valuable professional information and discover future business patterns and trends and multiple options and trends for customers. Big data has thus become the lifeblood of decision-making, the raw material of the liability process, and turned into knowledge that can be exploited in all areas of life.

The rapid growth in the production of data in terms of volume, source, velocity, and variety, faced great challenges in how to deal with these data and maximize the use of it so that, in the meantime, big data has become a recent event by most national, regional, and international bodies and institutions, and a wide range of research in database management, methodologies, and procedures that can be adopted to harness big data to serve the sustainable development goals.

*What is big data, and how can it be used to achieve the SDGs?*

To deal with this problem, we will address the following points:

- What is the meaning of "big data" and what are its characteristics?
- What is sustainable development, and what are its objectives?
- How can the opportunities presented by big data be used to serve the SDGs?
- What are the main challenges associated with the exploitation of big data?

To answer these questions, we divided our study into the following sections:

- **Section 19.2:** The concept of big data and its characteristics.
- **Section 19.3:** The concept of sustainable development and its objectives.
- **Section 19.4:** Opportunities provided by big data for sustainable development.
- **Section 19.5:** Big data exploitation challenges.

## 19.2  BIG DATA CONCEPT AND CHARACTERISTICS

Not so long ago, data was confined in a category of organized databases in folders, files, tables, and others. This now constitutes no more than 10% of the world's total information. However, access to new, accelerated, and unstructured information sources such as e-mails, videos, Facebook posts, tweets, WhatsApp chat messages, and other sources have set large databases.

### 19.2.1  BIG DATA CONCEPT

The definition of big data may not be clear because the volume of data is not defined by place or time. With the current acceleration in the development of information and communication technology (ICT), the huge data at present may not be huge in the future. In addition, a large amount of data for a particular person or institution may not be large for another person or institution. Below we will try to provide some definitions for big data.

In 2011, the Mackenzie World Institute launched a definition of big data: a data set larger than the ability of traditional databases to capture, store, manage, and analyze those data. In this context, the term "big data" in the field of information technology has been given to a set of very large and complex data packets that are difficult to deal with by traditional database management systems.

Also, the United Nations describes big data as "sources of data of large volumes, high velocity and data variety, which require new tools and methods for their capture, preservation, management, and effective processing" [1].

Big data is a set of large and complex data that has unique characteristics (such as volume, velocity, variety, data validity), which cannot be efficiently processed using existing and traditional technology to take advantage of it.

The challenges associated with this type of data are its provision, processing, storage, analysis, research, sharing, transmission, imaging, and updating, as well as the preservation of the specificities that accompany it [2].

## 19.2.2 CHARACTERISTICS

Characteristics of big data are as follows [3]:

1. **Volume:** Amount of data extracted from a source, which determines the value and volume of data to be classified as big data. By 2020, cyberspace will contain approximately 40,000 MB of data ready for analysis and debriefing.
2. **Variety:** It means the variety of the data extracted, which helps users, whether researchers or analysts, to choose the appropriate data for their field of research and includes structured and unstructured data in databases such as images, clips, and recordings of audio and video, SMS, and call logs and maps data (GPS) and requires time and effort to be configured in a suitable format for processing and analysis.
3. **Velocity:** The velocity of production and extraction of data to cover the demand for is the velocity of a critical element in the decision-making based on these data, which is the time between the moment of the arrival of this data and the moment of decision.
4. **Reliability and Argument:** It refers to the reliability of data source, accuracy, correctness, and timeliness of data, as one in three executives do not trust the data presented to them for decision. For example, there are studies that estimate that the damage to good data on the US economy is estimated annually at $3.1 trillion.

## 19.3   SUSTAINABLE DEVELOPMENT MEANING AND DIMENSIONS

The sustainable development idiom was first mentioned in the 1987 report of the World Commission on Environment and Development (WCED), which is defined in this report as: "that development which meets the needs of the present without compromising the ability of future generations to meet their needs."

In 1989, Barbier defined sustainable development more broadly as encompassing the creation of a social and economic system that would ensure support for the following objectives: an increase in real income, an improvement in education level, and an improvement in population health [4].

The United Nations has set seventeen development goals as a plan for a better future for all. These goals face current global challenges [5]:

1. Elimination of poverty;
2. Elimination of hunger;
3. Good health and well-being;
4. Good education;
5. Gender equality;
6. Clean water and hygiene;
7. Clean and affordable energy;
8. Decent work and economic growth;
9. Industry, innovation, and infrastructure;
10. Reduce inequalities;
11. Lasting cities and communities;
12. Responsible consumption and production;
13. Climate action;
14. Life underwater;
15. Wildlife;
16. Peace, justice, and strong institutions;
17. Partnerships to achieve goals.

## 19.4 OPPORTUNITIES PROVIDED BY BIG DATA FOR SUSTAINABLE DEVELOPMENT

Big data will contribute to the achievement of sustainable development goals, especially in light of the large spread of the Internet. This new technology will contribute to the management of difficult issues at the local and global levels and help address them.

Big data technology can help promote enterprise development. It allows the development of tailored and accurate analysis, to businesses, for existing and potential customers, improves user experience, addresses manufacturing inefficiencies, and associated processes [6].

In the health sector, big data technology may allow improved health care to diagnose treatment, collect patient data beyond what it exchanges with the doctor on sporadic visits; detect early disease progression and proactively treat it at the individual and community level, and arrive at more effective treatments for a range of diseases. In particular, data maps can help support response to disease outbreaks.

In agriculture, big data technology opens up new opportunities in agriculture, including food security.

Big data technologies can balance energy supply and demand by installing smart grids that enhance the role that renewable energy sources play in the distribution and the production of energy by supplying households with solar panels or wind turbines to feed the power grid with surplus energy.

Real-time information provided by smart grids helps electrical supply companies improve their responses to changes in demand, power supply costs, and emissions, as well as to avoid power outages.

Efficient water production and distribution, especially in urban areas, is an ever-present challenge for governments. In this context, Internet-related tools, such as sensors, meters, and mobile phones, can be provided with functions that allow monitoring and study of water quality for smarter water management, as in the case of a wireless sensor network.

The collection and measurement of development indicators will be central to monitoring progress in the achievement of development goals. In this context, stakeholders, including international organizations, academics, and corporations, are seeking to explore ways in which big data can contribute to the objectives of monitoring and economic policy-making activities.

Big data can provide a unique service to educational institutions if they can use, process, store, and manage it. It provides a better understanding of the beneficiaries and their requirements and helps to make appropriate and rational decisions within these institutions in a more efficient and effective manner using the information extracted from the beneficiary databases and thus reduce costs for students and faculty [7].

Big data and Internet sensors for research and development enable researchers to analyze and discover patterns of scientific data that were not available until recently. There are many areas in which such capabilities are of great benefits, such as meteorological forecasts and the human mind's exploration.

Many big data technologies and artificial intelligence algorithms are based on open-source technology. This makes them freely available for use, exchange, change, and adaptation by creating opportunities for local and pro-poor innovation adapted to regional needs and markets.

Using open licenses by colleges and universities can develop versions of big data technologies and machine learning algorithms that address local challenges. However, working with these technologies and innovating through it requires appropriate skills (such as the ability to analyze and preview big data, requiring mathematical and computational skills), and thus highlights the importance of capacity building to take advantage of these new technologies [6].

## 19.5  ISSUES RELATED TO BIG DATA

Governments and institutions face a number of greetings and risks in their implementation of large data projects. Concerns include legal issues related to the privacy and the access to data, lack of human capacity, and the expansion of IT infrastructure to address the demands of large data sets, including challenges we find:

1. **Privacy:** Tracking personal issues and monitoring to analyze (web) page or social network visits, phone calls, and e-mail, and track and monitor religious, political, or terrorist tendencies is a risk of big data, for example, former US President Obama and former British Prime Minister Cameron (such as Google and Facebook) in collaboration with intelligence in tracking terrorists on social networks and the Internet.

   Furthermore, this action provoked protests from human rights organizations, which represented a violation of personal privacy. Big data includes collecting and analyzing personal data about individuals, population information, business, government, and military activities, water consumption, energy, national reports for various purposes, online IP abuse, social media, e-mail, and free subscriptions to websites.

   Big data technologies and services tax the protection of individuals' privacy and sensitive data during the processing cycle, while keeping that data stored, and scalability is a real threat to security and privacy.

   The biggest obstacle to the manipulation of big data in criminal activity prediction is that programmers and law enforcement are not cooperating [8].

Another challenge is determining what to do when data analysis indicates that a person is about to commit a crime. Prosecutors can ask a judge to place someone under house arrest or imprisonment if there is sufficient physical evidence, but arrest a person based on big data analysis can be more difficult to convince a judge.

Perhaps data and software do not always show the full picture, although big data programs and their associated technologies provide data and information in advance, contribute to law enforcement and crime prevention, but before we can do that, programs need to be improved and answer important questions, such as effects on personal privacy.

2.  **Data Coverage Challenges:** Often, there are categories of the target community for which no data are available, or data may be available, but incomplete, as the data may not cover certain variables under consideration.

    This requires the availability of expertise, technical, and statistical capabilities in addition to tools and software for data processing and compensation for missing values and testing the quality of statistical data as a final output.

3.  **Big Data Volume:** Dealing with the large volume of data is a major challenge for all institutions in most countries. This challenge concerns the ability of these institutions to store a large amount of data, as well as procedures for processing data and minimizing the impact of expected procedural errors on results [1].

4.  **Data Usage Laws and Regulations:** Analysis of the huge data is based on data that can be discovered, accessed, and used, and data must be collected, stored, used, and managed in conformity with the relevant laws and regulations.

    To realize the full potential of big data, departments and government agencies need to focus their attention on making as much data as possible open and accessible. States are currently publishing an open data policy to facilitate the use of government data within government agencies, as well as by the public [9].

    Data stored in one country may be of utmost importance to institutions in another country, so the fact that different countries have adopted different laws and regulations on data storage, how they are used, and what data is accessible is a big obstacle for users of big data.

5. **Data Access:** Although the majority of new technological transformations are seen as a powerful tool that may, in different ways, help the political, economic, and social empowerment of the average user, big data sensors are often in the hands of intermediate institutions of government agencies and large corporations and do not have ordinary people causing disparities of power. Between those who have the data and those who don't [10].

   Big data is also expected to open a new front of disparity of forces among nations of the world, dividing them between those who know and those who do not. The coming wars will be essentially information wars, so who will have access to and analyze big data will be able to provide new opportunities to enhance his intelligence and take the appropriate decision for his national security.

6. **Market Access:** Increased dissemination and the exploitation of big data technologies can create highly diverse jobs. For instance, in the United States, there are estimated to be about 500,000 jobs in the big data sector in 2014.

   Countries' ability to participate competitively and actively in global markets depends on, inter alia, their ability to create a trained workforce capable of understanding the unprecedented flow of data generated by these innovations and harnessing them to produce real value.

   Some professions related to big data include mathematics, computing, and engineering [6]. Moreover, than a trained workforce, big data's effective application requires a wide range of supporting infrastructure and enabling policy frameworks, such as cloud computing resources and interoperability standards.

7. **Data Management, Use, and Analysis:** Large businesses are looking for the right ways to store, manage, use, and analyze data to make the most of it. A New Vantage Partners survey found that only about (37.1%) of businesses believe that they are successful in trying to use big data, while (71.7%) of institutions believe that they didn't formulate a culture of data dependence yet, and the rate of (53.1%) of institutions stated that they didn't yet deal with data as a commercial asset.

   These institutions often fail to know the basics about what big data is, what its benefits are, what infrastructure to adopt, and so on. Without a deep understanding of all these fundamentals, the big data

Adoption Project will fail, and organizations may waste a lot of time and resources on things that employees do not know how to use [11].

If employees are unaware of the benefits of big data or do not wish to change the methodology of existing processes for adoption, they will resist it, thereby impeding the institution's progress.

8. **Data Quality:** The integration of the data is a problem that organizations face because the data they need to use comes from a variety of sources. For example, e-commerce companies used to analyze data from website records, call centers, and competitors' websites, and there is obviously data inconsistency, so hard to match.

   A bigger challenge is unreliable data, since big data is not accurate, not only because it can contain wrong information, but because it can be repetitive, as well as may contain discrepancies. Poor quality data are unlikely to provide any useful information or important opportunities, and inaccurate information may increase the risk of making wrong decisions.

9. **Big Money Expenditure:** Big data accreditation projects involve a lot of expenses, taking into account the costs of new hardware, hiring staff such as system administrators, developers, etc. Although the necessary systems are open source, there is always a need to pay for new software development, preparation, and maintenance. If one decides to rely on a large cloud-based data solution, organizations will still need to hire staff, pay for cloud services, develop big data solutions, and set up and maintain frameworks.

10. **Timing:** The accessibility and timeliness of finding items in a limited time in a large database is another new challenge in data processing, and the ability, for new types of criteria, to respond to data requests with limited times is an additional challenge.

11. **Financial Loss and Reputation:** This is a result of big data penetration.

## 19.6  CONCLUSION

Through the above study, it becomes clear that big data projects can play a vital role in economic growth and sustainable development by their exploitation to improve health services, open learning, increase access to education, stimulate innovation and agriculture and business, and open the door to increased transparency, justice, and efficiency in service delivery and access to energy sources.

However, the barriers to harnessing big data as a useful tool for devel-opment are many, but not insurmountable, as big data bring the expecting results. In the coming years, if conditions are right to exploit them, by (I) making greater efforts to improve affordable access to ICTs and data whether in urban and rural communities, (ii) promoting the continuous development of network security and privacy, (iii) opening learning and lifelong learning opportunities for all members of society, and (iv) building human capacity in the exploitation of big data, etc.

## KEYWORDS

- **big data**
- **privacy**
- **sustainable development**

## REFERENCES

1. Statistics Centre. *General Concepts about Big Data, Methodological and Quality Guides.* Retrieved from: https://www.scad.gov.ae/MethodologyDocumentLib/13%20 %D8%AF%D9%84%D9%8A%D9%84%20%D8%A7%D9%84%D8%A8%D9%8A %D8%A7%D9%86%D8%A7%D8%AA%20%D8%A7%D9%84%D9%83%D8%A8 %D9%8A%D8%B1%D8%A9%20%D8%A7%D9%84%D9%86%D8%B3%D8%AE %D8%A9%20%D8%A7%D9%84%D9%86%D9%87%D8%A7%D8%A6%D9%8A% D8%A9.pdf (accessed on 22 October 2020).
2. Big Data, (2019). *Free Encyclopedia Wikipedia.* Retrieved from: https://en.wikipedia. org/wiki (accessed on 22 October 2020).
3. Big Data in Arabic, (2014). *Big Data Opportunities and Challenges.* Retrieved from: https://bigdatainarabic.wordpress.com (accessed on 22 October 2020).
4. Moetasim, M. I., (2015). *The Role of Investments in Achieving Sustainable Development.* (Syria as a Model), PhD Thesis in Economic Sciences, Damascus University.
5. United Nations Development Program, (2018). *Sustainable Development Goals.*
6. United Nations, (2016). *Economic and Social Council, Commission on Science and Technology for Development, Outlook for Digital Development.* Retrieved from: https:// unctad.org/meetings/en/SessionalDocuments/ecn162016d3_en.pdf (accessed on 22 October 2020).
7. Al-Jaid, A. R. A., (2017). *Big Data Analysis and Education Improvement.* Retrieved from: https://www.new-educ.com/ (accessed on 22 October 2020).
8. Abu, B. S. A., (2017). *Big Data: Characteristics, Opportunities, and Power.* Retrieved from: https://mail-attachment.googleusercontent.com/attachment/u/0/?u i=2&ik=767b6df1f5&attid=0.2&permmsgid=msg-f:1683924430273232275&th=

175e8058f709bd93&view=att&disp=inline&saddbat=ANGjdJ_-IJyWs8zRQlqTV
Xc60qykZrFTcVh54vnXckJj1eYP0iqyxVFUK5lEY9SSVn0vDhpaqO8AXi66rW
CET1WVxeZqwz3sbftSgKw6pOSlGJsZo2cukjD0nsseV7pOpcjXsUaEvxzsIgUM
1TfK7-Ma2dW4iIbt1-ZumX4NGoDvloHvGwMQr_wXdBOBPGQrgC9x-_Kfo-
WN921bsEhYrrlKaK3STsKwv8XNZyicGZyUZCGY6JtpA3QEKjPh7M3ocPc5
Z5KDnFjjjPow9aHeNkIery2puC2YJjbbcO1xtg--RSs233TaJpGZVJ8PdwNR2J-
8llfhAGqR5PrA7SFcckSjL3JcAqZsUqDMkYIA2uvz7aKX_KFl5dvgmPWD263mP_
cJ9_UKkHUk0-5UiRSEDvyq_pcf_NXY8x-xZYXulAPjc3H-HjWhvZY_XrwI_weRv
yJYUdILEUFunzOupsN1cfzfyJxworgdp4O1ZroC9rQbR8bg1a1HDLtsRkE7b2SPz
9yr9xX6qaGef3Rxw4ZSI9Wy34BQssvudwmr67ZAYEr0o6UFKVkvkvFUxK0MLo
1UKv-juRfwwv8SzmnXKYOwlnfKdGP7P-99CcxezpBvqALtr1z5CXbZ3g_0whwpzs
J7BGRg00pfweIfo9u4_ICMyebEyCMrZn-JcGVGgIOtRBIuHQBteZMlof5lEfKiAL8
(accessed on 22 October 2020).

9. Ministry of Information and Communication Technology, (2014). *Big Data, Balancing Benefits and Risks* (p. 08). Qatar. Retrieved from: http://www.motc.gov.qa/sites/default/files/lbynt_ldkhm_thqyq_ltwzn_byn_lmzy_wlmkhtr.pdf (accessed on 22 October 2020).

10. Islam, H., (2018). The new knowledge oil. *On the Concept of Big Data and its Future Challenges*. The Arab Thought Foundation, Retrieved from: https://arabthought.org/en/researchcenter/ofoqelectronic-article-details?id=905, http://www.arabstates.undp.org/content/rbas/en/home/sustainable-development-goals.html (accessed on 22 October 2020).

11. Aitnews, T. V., (2019). *The Five Most Important Challenges Faced by Companies When Using Big Data*. The Arab portal for technical news, Retrieved from: https://aitnews.com/2019/05/3 (accessed on 22 October 2020).

# CHAPTER 20

# Big Data for Sustainable Development Goals: Theoretical Approach

FATIMA LALMI[1] and RAFIKA BENAICHOUBA[2]

[1]PhD in Economic Sciences, Senior Lecturer, University of Abdelhamid Ibn Badis, Mostaganem, Algeria, E-mail: lalmi.fatima@yahoo.fr

[2]PhD in Economic Sciences, Senior Lecturer, University of Djillali Bounaama, Khemis Maliana, Algeria, E-mail: benaichoubarafika@yahoo.fr

## ABSTRACT

Big data applications became increasingly important in different fields. This study aims to show the role of big data in achieving sustainable development purposes, using documentary and literary sources for gathering and analyzing information about the topic. This study reached several results, mainly: big data provides opportunities in sustainable development like improvement of different services levels, promotion of innovation, and motivation of investment and creation of new sustainable jobs, improvement of institutions' performance, and facilitation of making a decision.

## 20.1  INTRODUCTION

The unsustainable practices based on increased consumption of non-renewable resources to achieve economic purposes led to increasing negative environmental impacts such as pollution and climate change. In this circumstance, global orientation has been made towards sustainable development based on economic, social, and environmental dimensions. However, achieving sustainable development goals requires the availability of various data and information for the formulation and follow-up of effective policies.

Recently, by observing the development of data, we found that a tremendous amount of data produced, stored, and made available from multiples sites has become a force source of any society based on knowledge, as well, the capacity to address data and complex analyses has also increased to obtain important information.

Thereby, big data is becoming a very famous concept in academia. It has been used in many fields of modern society and is widely used in banking, education, logistics, and other fields [1].

Its applications offer the ability for collecting and analyzing information in real-time from different states for policies related to the 2030 Agenda's 17 goals and their 169 targets and help decision-makers to devise strategies in order to improve the economies of societies, achieve competitiveness, preserve the environment and health, protect the community and meet the needs and improve living standards [2].

Hence, big data is needed to help governments make effective progress at various levels to become sustainable [3]. Therefore, in this context, the purpose of this research is to demonstrate the importance of big data and its positive effects on the economy, society, and environment by focusing on its role as a tool for policy planning and sustainable development.

The problem of this research steams from the following question:

*How can Big Data contribute to achieving sustainable development goals in light of Big Data applications' opportunities for development planning and evaluations?*

The chapter is structured as Section 20.2 contains big data definition, characteristics, dimensions, and sources. Section 20.2.2 gives information about the sustainable development concept and its pillars and goals. Section 20.3 discusses the relationship between big data and sustainable development goals. Section 20.4 is the conclusion.

## 20.2   THEORETICAL BACKGROUND

### 20.2.1   BIG DATA: DEFINITION, CHARACTERISTICS, DIMENSIONS, AND SOURCES

1. **Definition and Characteristics:** There are various definitions of big data available in the literature. Among them is one from Gartner in 2012, defined as: "big data is high volume, high velocity, and high variety information assets that require new forms of processing to

enable enhanced decision making, insight discovery, and process optimization" [4].

This definition refers to the three basic characteristics of big data, also known as the 3V's:

- Volume refers to the large scale of big data, which requires innovative tools for collection, storage, and analysis.
- Velocity represents the speed at which data is created or updated, pointing to the real-time nature of big data.
- Variety signifies the variation in types of data. Big data comes in diverse and dissimilar forms from multiple sources.

After a few years, IBM added another characteristic or "V" on the top of Laney's 3V's notation, which is known as 4V's of big data. It describes each "V" as following [5]:

i. The volume stands for the scale of data.
ii. Velocity denotes the speed of data transfers.
iii. Variety means different forms of data.
iv. Veracity indicates the uncertainty of data.

In 2013, Yuri Demchenko extended this 4V model to a 5V model by including the value dimension, which refers to the process of extracting valuable information from large sets of social data, and it is usually referred to as big data analytics [6]. It means the core of big data that address the cost/benefit proposition.

Other organizations and big data practitioners proposed a 6V's model by adding: Veracity, Variability, and Visibility. The first V, Veracity, means the degree of reliability and credibility of data sources. The second V, Variability, refers to the complexity of the data set. The third V, Visibility, emphasizes the data accuracy in order to make a decision [5].

All the above definitions agree that big data is a large volume of high velocity, complex, variable, and visible data that require advanced techniques to enable the collection, storage, distribution, management, and analysis of information to facilitate decision-making.

Hence, big data aims to gain hindsight from historical data, insight from understanding the issues, and foresight by forecasting in the future in a cost-effective manner.

Moreover, Timo Elliott stated that each definition of big data focuses on a particular issue from one aspect of big data, so he classified these definitions into seven types, shown in Table 20.1 [5].

**TABLE 20.1**   Seven Types of Big Data Definitions by T. Elliott

| No. | Types | Description |
|-----|-------|-------------|
| 1. | The original big data (3 V's) | The original type of definition is referred to Douglas Laney's volume, velocity, and variety, or 3 V's, such as 4 V's, 5 V's, up to 6 V's |
| 2. | Big data as technology | This type of definition is driven by new technology development, such as MapReduce, bulk synchronous parallel (BSP_Hama), resilient distributed datasets (RDD, Spark), and Lambda architecture (Flink) |
| 3. | Big data as an application | This kind of definition emphasizes different applications based on different types of big data. Barry Devin defined it as an application of process-mediated data, human-sourced information, and machine-generated data. Shaun Connolly focused on analyzing transactions, interactions, and observation of data. It looks for hindsight of data. |
| 4. | Big data as signals | This is another type of application-oriented definition, but it focuses on timing rather than a data type. It looks for the foresight of data or a new "signal" pattern in a dataset. |
| 5. | Big data as an opportunity | Matt Aslett: "Big data as analyzing data that was previously ignored because of technology limitations." It highlights many potential opportunities by digging-in the collected or achieved datasets when new technologies are variable. |
| 6. | Big data as a metaphor | It defines big data as a human thinking process. It elevates BDA to a new level, which means BDS is not a type of analytic tool; it's actually an extension of the human brain. |
| 7. | Big data as a new term for old stuff | This definition simply means the new bottle (relabel the new term "big data") for old wine (BI, data mining (DM), or other traditional data analytic activities). It is one of the most cynical ways to define big data. |

*Source*: Adapted from Ref. [5].

2.  **Big Data Dimensions:** Big data as an integrated approach for development includes three dimensions, shown in Figure 20.1 [7]:

- Data generation generates and collects large volumes of data using smart technologies.

- Data analytics involves the organization and integration of various sources of data and the identification of previous patterns and associations in data, explained in Figure 20.1.
- Data ecosystem involves producers, analysts, regulators, and users of big data to combine big data analytics based on quantitative and qualitative analysis to ensure successful applications of big data, including interactions between humans and big data digital technology.

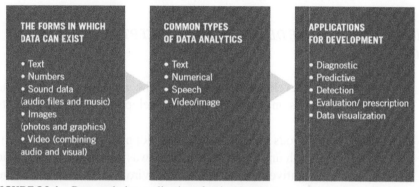

**FIGURE 20.1**   Data analytics applications for development.

*Source*: Reprinted from Ref. [14]. © UN Global Pulse, 2016

3. **Big Data Sources:** big data has many sources such as social media, machine-generated data, sensor data, transaction data, and the Internet of things (IoT), which can be summarized as follows [8]:
   - Social media is a web-based service that allows individuals to construct a public or semipublic profile within a bounded system, communicate with other users, and view the pages and details provided by other users within the system. This data source contains a lot of information which is generated using URL (Uniform resource language) to share or exchange information in virtual communities and network for an example: Facebook, Twitter, LinkedIn, and YouTube.
   - Machine-generated data information is automatically generated from both hardware and software, such as smart meters that continuously stream data about electricity, water, or gas consumption.
   - Sensor data are collected from various sensing devices, and these are used to measure physical quantities. There are two types of

sensor; the first one is fixed sensors like weather sensors, traffic sensors, and scientific sensors. The second one is mobile sensors such as mobile phone location (GPS) and satellite images.

- Transaction data involves a time dimension to illustrate the data, for example, commercial transactions.
- Internet of things generates a huge amount of data from a lot of internet-connected devices such as smartphones and digital cameras.

### 20.2.2　SUSTAINABLE DEVELOPMENT: CONCEPT, PILLARS, AND GOALS

1. **Sustainable Development Concept:** The concept of sustainable development has undergone different developmental phases since its introduction.

   Previous periods considered the development as a synonym for economic growth until the 1970s when it was obvious that economic growth has negative influences on the environment, such as climate changes, ecosystem disturbances, and natural disasters.

   At the same time, the increasing consumption of non-renewable resources, especially the stock of fossil fuels, led to the deliberation of the needs of future generations and created a prerequisite for describing the attitude of long-term and rational use of non-renewable resources [9].

   Under these circumstances, a group of scientists, economists, and humanists from ten countries gathered as an independent organization named "Roman Club" in Rome in 1968 in order to discuss the current problems and future challenges of humankind, including limited natural resources, environmental degradation, population growth, etc.

   This club has published two editions, the first one titled: Limits of Growth in 1972, and the second one titled: Mankind at the Turning Point in 1974, containing the results of their research and warning that excessive industrialization and economic development would soon cross the ecological boundaries [10].

   Different organizations participated in creating the concept of sustainable development; in particular, United Nations – which was established in 1945 – actively acted in this field by organizing many conferences and publishing various reports for achieving sustainable development goals.

Since the introduction of the concept, a lot of conferences, meetings, and congresses have been held, resulting in various reports, conventions, agreements, and dealing with the environmental problems, summarized in Table 20.2 [9].

**TABLE 20.2**  Overview of Important Activities Related to the Concept of Sustainable Development

| Year | Activities | Description |
|------|-----------|-------------|
| 1969 | UN published the report titled: Man, and His Environment | Focusing on Avoiding global environmental degradation. |
| 1972 | First UN world conference about the human environment, Stockholm | Setting up a global environmental framework, known as the Belgrade Charter. |
| 1975 | International Congress of the Human Environment, Kyoto, Japan | Emphasizes Problems discussed earlier in Stockholm in 1972. |
| 1979 | The First World Climate Conference, Geneva | Focused on the creation of climate change research and program monitoring. |
| 1981 | The First UN Conference on Least Developed Countries, Paris | Resulted in a report containing guidelines and measures for helping the underdeveloped countries. |
| 1984 | Establishment of the United Nations World Commission on Environment and Development (WCED) | The commission's task was the cooperation between developed/developing countries and the adaptation of global development plans on environmental conservation. |
| 1987 | WCED report called: Our Common Future | A report contained the fundamental principles of the concept of sustainable development. |
| 1990 | The Second World Climate Conference, Geneva | Focused on the further development of climate change and the creation of the Global Climate Change Monitoring System. |
| 1992 | UN conference on Environment and Development, Rio de Janeiro | The Rio Declaration and Agenda 21 created an Action Plan principle of sustainable development and the framework for future tasks. |

**TABLE 20.2**   *(Continued)*

| Year | Activities | Description |
|------|-----------|-------------|
| 1997 | Kyoto Climate Change Conference, Kyoto, Japan | The Kyoto Protocol was signed between countries in order to reduce $CO_2$ and other greenhouse gas emissions, with commencement in 2005. |
| 2000 | UN published Millennium Declaration | Declaration containing eight Millennium goals targeted by 2015. |
| 2002 | The World Summit on Sustainable Development, Johannesburg | The report contained the results achieved during the time from the Rio Conference and set the guidelines for implementation of the concept in the future. |
| 2009 | The Third World Climate Conference, Geneva | Further development of the Global Climate Change Monitoring System with the aim of timely anticipation of possible disasters. |
| 2012 | UN conference Rio+20, Rio de Janeiro | Twenty years from the Rio conference resulted in a report: The Future we want to be renewed – the commitment to the goals of sustainable development and encouraged issues of the global green economy. |
| 2015 | UN Sustainable Development Summit, New York | The UN 2030 Agenda for Sustainable Development was established, setting up 17 Millennium Development Goals, which should be achieved by 2030. |
| 2015 | UN conference on Climate Change COP 21, Paris | Agreement on the reduction of greenhouse gases in order to limit global warming. |

*Source*: Reprinted from Klarin, T. 2018. [Ref. 9].

Among various activities cited in the previous table, the history of the concept of sustainable development is divided into three periods. The first one covers the period from economic theories to the First UN Conference on the Human Environment held in Stockholm in 1972, which marked the introduction of the concept of sustainable development without a full association between environmental problems and economic development [9].

The second period extended between 1973 and 1990, marked by holding several conferences and many issued reports, especially in 1987, the United Nations World Commission on Environment and Development (WCED) published a report Our Common Future contained the fundamental principles of the concept of sustainable development, based on analysis of the conditions in the world (socio-economic development) and order, environmental degradation, population growth, poverty, etc.

In this period, sustainable development is defined as "development that meets the needs of the present without compromising the ability of future generations to meet their own needs" [11]. The third period starts after a previous period and lasts until nowadays; it contains important events such as United Nations Conference on Environment and Development held in Rio de Janeiro in 1992, resulted in the following documents [9]:

- Rio declaration on environment and development, which contains 27-principles of sustainable development, forming the basis for future policy, decision making, and balance between socio-economic development and the environment.
- Agenda 21: a global program with the aims of sustainable development, action plans, and resources for their implementation. It provides guidelines for socio-economic development in line with environmental conservation.

From these documents, the three fundamental elements of sustainable development are identified:

- **First Element:** The concept of development based on socio-economic development in line with environmental constraints.
- **Second Element:** The concept of needs.
- **Third Element:** The concept of the future generation.

2. **Sustainable Development Pillars:** Sustainable development based on balancing the three pillars of sustainability: the first is ecological sustainability, which means to maintain the quality of the environment needed for economic activities and quality of life; the second is social sustainability based on preservation of society and cultural identity, etc. The third is economic sustainability signifies maintaining the various sources (natural, human, social) to achieve income and living standards. The balance between these pillars is

difficult to achieve because each one in the process of achieving its goals, must respect the other pillars' interests [12].

3. **Sustainable Development Goals:** In 2015, the United Nations announced 17 sustainable goals which should be achieved by 2030; we can summarize most of them as follows: eradicating poverty and hunger, good health and education, clean energy and water, ensuring gender equality, ensuring environmental sustainability and global partnership for development [13].

## 20.3  RESULTS AND DISCUSSION

### 20.3.1  BIG DATA AND SUSTAINABLE DEVELOPMENT GOALS

Big data plays an active role in achieving sustainable development goals by the improvement of understanding issues or problems and offers policy-making support for the development in three main ways: the first way is an early warning, which means early detection of anomalies to enable faster responses to populations in crisis times.

The second way is real-time awareness; signifies fine-grained representation of reality through big data that can inform the design and targeting of programs and policies. The third way is real-time feedback; concerns adjustments that can be possible by real-time monitoring the impact of programs and policies [14].

Big data can improve current official statistical systems in many ways, such as: providing complementary statistical information in the same statistical domain but from other perspectives, improving estimation from statistical sources, and providing new statistical information in a particular statistical domain [15].

Moreover, big data contributes to the management of aerial issues at the local, national, and international levels through their applications in enterprise development and in various economic sectors such as health, agriculture, energy, and water resources, which can be explained as follows:

- In enterprise development, big data enables these enterprises to develop accurate analyses of current and potential clients, which could address a lack of efficiency in industrialization. For example, the first product of microinsurance has been established and distributed in Africa using a mobile phone network.

- In the health sector, big data can contribute to improving health care through the provision of various data about patients and its illness from which can detect early evolution and treatment in time. Data maps also support the response to the spread of diseases. For example, the Ugandan Ministry of Health used data maps in the spread of typhoid disease to facilitate the decision concerning the allocation of medicines and the distribution of health teams.
- In the agriculture sector, big data applications for agricultural crops contribute to achieving food security and eliminating hunger. For example, the implementation of software produced by the Indian company CropIn has facilitated crop management by providing data on crop growth at various stages of production.
- In the energy sector, big data applications contribute to the balance between demand and supply of energy through the provision of available data, which enable energy institutions to better respond to demand changes, supply costs of electricity, avoid power outages as well as the exploitation of renewable energy to meet of increasing demand for electricity.
- In the water field, big data applications facilitate the management of water resources. For example, AQUASTAT is FAO's global water system, which collects and analyses data on water resources, water uses, and other information [7].

Furthermore, the United Nations published a report about how big data applications can be used to achieve sustainable development, shown in Figure 20.2 [13].

**FIGURE 20.2**  Big data applications for sustainable development.

## 20.4  CONCLUSION

This study tries to give an overview of big data characteristics, dimensions, sources. We also tried to show the essential role of big data applications in the improvement of decision making and achieving sustainable development goals, using enormous reports and studies in order to illustrate how big data is currently being used to develop help for disaster relief, mine citizen feedback and map population movement to support response to crises outbreaks.

Big data applications provide many opportunities in achieving sustainable development goals. The study reached several results, which can be summarized as follows:

- Providing accurate data to improve the level of different services as best health care, clean energy.
- Usage of data in different economic fields as using databases in investment motivation and creating jobs, thereby increasing the efficiency and quality of services and goods.
- Promote innovation by using big databases in detailed studies to innovate new services contributing to improving institutions' performance.

Big data applications face several challenges, such as a lack of human resources competencies, which enable the optimization benefits of using big data, its privacy, and security.

### KEYWORDS

- **big data**
- **sustainable development**
- **goals**
- **decisions-making**

### REFERENCES

1. Wang, Z., & Xue, M., (2019). Big data: New tend to sustainable consumption research. *Journal of Cleaner Production, 236,* 1–9.
2. Singh, S. K., & Abdulnasser, E. K., (2018). Role of big data analytics in developing sustainable capabilities. *Journal of Cleaner Production, 213,* 1264–1273.

3. Song, M., Cen, L., Zheng, Z., Fisher, R., Liang, X., Wang, Y., & Huisingh, D., (2016). How would big data support societal development and environmental sustainability? Insights and practices. *Journal of Cleaner Production, 142*, 489–500.

4. Abdulla, M.A., & Manners, L.C., Louis, K. M., (2018). The role of information centers and managing big data in Saudi Arabia in support of sustainable development. *24th Annual Conference and Exhibition of the SLA/AGC: Big Data and its Investment Prospects* (pp. 1–38). Muscat.

5. Buyya, R., Calheiros, R., & Pastjerdi, A. (2016). *Big Data: Principles and Paradigms* (p. 494). Morgan Kaufmann Edition; USA.

6. Koseleva, N., & Ropaite, G., (2017). Big Data in building efficiency: Understanding of bid data and main challenges. *Procedia Engineering, 172*, 544–549.

7. United Nations Global Pulse. *Integration of Big Data into the Monitoring and Evaluation of Development Programs* (p. 73). New York. https://www.unglobalpulse.org/big-data-monitoring-and-evaluation-report (accessed on 22 October 2020).

8. Boyd, M. D., & Ellison, B. N., (2007). Social network sites: Definition, history, and scholarship. *Journal of Computer-Mediated Communication, 13*(1), 210–230.

9. Klarin, T. (2018). The Concept of Sustainable Development: From its Beginning to the Contemporary Issues, Zagreb International Review of Economics and Business, 21(1), 67-94. doi: https://doi.org/10.2478/zireb-2018-0005.

10. Meadows, D., Randers, J. & Meadows, D., (2004). *Limits to Growth: The 30-Year Update* (p. 218). Earthscan Edition, UK.

11. United Nations WCED, (1987). *Our Common Future* (p. 247). United Nations, New York. https://sustainabledevelopment.un.org/content/documents/5987our-common-future.pdf (accessed on 22 October 2020).

12. Strange, T., & Bayley, A., (2008). *Sustainable Development: Linking Economy, Society, and Environment* (p. 82). OCDE insights series.

13. United Nations, (2017). *The Sustainable Development Goals Report* (p. 64). New York. https://unstats.un.org/sdgs/files/report/2017/thesustainabledevelopmentgoalsreport 2017.pdf (accessed on 22 October 2020).

14. UN Global Pulse, 'Integrating Big Data into the Monitoring and Evaluation of Development Programmes,' 2016. https://www.unglobalpulse.org/wp-content/uploads/2016/12/integratingbigdataintomedpwebungp-161213223139.pdf

15. Sriganesh, L., Thavisha, P. G., & Shazna, Z., (2017). *Mapping Big Data Solutions for the Sustainable Development Goals* (p. 81). IDRC Edition, The international development research center, Canada.

# CHAPTER 21

# Using Big Data in Official Statistics for Sustainable Development

KHADRA RACHEDI[1] and FATIMA RACHEDI[2]

[1]University of Oran 2, Algeria, E-mail: Kha-dra@hotmail.fr

[2]University of Larbi Ben Mhidi, Oum El Bouagui, Algeria,
E-mail: rachedi.fatima@yahoo.fr

## ABSTRACT

A large amount of data is characterized by large size, diversity, and speed, making it an important source of official statistics, one of whose objectives is to monitor progress in achieving the goals of sustainable development. This chapter will try to address the importance of large data in improving and updating the statistical production process to strengthen the development agenda of sustainable development and highlight some models in countries and the large data sources used.

## 21.1 INTRODUCTION

The statistical data are important sources for studying different phenomena and making critical planning decisions. This requires sufficient statistical data that allow conducting social and economic studies of a purposeful scientific dimension, where they are often derived from census, demographic surveys, and other traditional sources that play a major role in achieving sustainable development goals.

Thanks to the immense development of technology, data have witnessed a great revolution that its weight manifested in the emergence of the "big data," which do not stop accumulation and diversity. As one of the most recent scientific disciplines, analyzing big data contributed to the development of many areas like health, marketing, and security due to many developed programs and applications that allow using them.

Based on the previous discussion, we proposed the following question:

*How do these data contribute to developing the official statistics for sustainable development?*

## 21.2   BASIC CONCEPTS

1. **Data:** It is the raw material for making data, and these data are in the form of numbers or images or pictures or non-titled or organized texts what make them difficult to be examined and analyzed.
2. **Big Data:** This concept demonstrates the accumulation of the increasing sources of information, and it's analyzing is huge in a way that is beyond the capacities of storage and analysis that were provided by first equipment and programs, what was achieved by the development of storing data capacity, and widening the scope of available data [1]. These are the results of major social and technological changes.

Generally, scientists agreed on four types of big data. We summarize them in the following:

- Structural data: This includes numbers and facts, easy to classification, and analysis.
- Video files, documents, and what is alike: (nonstructural) difficult in the analysis as a subject of its owners' points of view, so they are subjective.
- Data produced by digital sensors.
- Another type described by specialists "the big data of administrative registrations, digital books, and weather." They are of a great size what makes them classified within this type of data.

    Big data have characteristics known by 3V: the big volume, the flow velocity, and variety, in addition to other characteristics: value and veracity, so we have the 5V.

3. **Official Statistics:** They are the set of statics produced by governmental institutions and recognized international organizations, and they are associated with all general life aspects such as demographic and health statics, transportation, and agriculture.

Their main sources are the demographic and economic census, surveys by sample and administrative records, and others. These help to provide adequate indicators that reflect how much development programs are advanced, and they help to identify future needs.

4. **Sustainable Development:** UN defines sustainable development as the development that can meet the present's needs without compromising the ability of the next generations to reach their own needs. Sustainable development calls for corporate efforts for building the future for people where the earth is for all, sustainable, and resilient [2].

   For achieving sustainable development, it has to succeed between three elementary points: economic growth, social integration, and protecting the environment. These three elements are interrelated and all critical for the welfare of individuals and societies.

5. **Sustainable Developments Objectives:** It is to call all poor, rich, and middle-income economies to work for enhancing prosperity regarding the protection of the earth. These objectives are recognized by the fact that poverty eradication has to be linked to strategies that build economic growth. It also deals with social needs, including education, health, social security, job opportunities, and treating climate change and environment protection.

6. **Monitoring Sustainable Development Objectives:** This is done using a set of indicators that allow measuring the achievement of 17 objectives designed by the member nations in the UN, 2015, which can be accessed only by statistical data.

The objectives are:

- Poverty eradication;
- Eliminating famines and hunger totally;
- Good health and welfare;
- Good education;
- Gender equality;
- Healthy water and health facilities;
- Industry, innovation, and infrastructures;
- Reducing contrast;
- Sustainable local cities and societies;
- Responsible consumption and production;
- Climatic works;

- Life underwater;
- Life on land;
- Peace, work, and strong institutions;
- Partnership to achieve goals.


## 21.3  BIG DATA IN OFFICIAL STATICS

Official statics that allows the right decision-makings cannot be built except for correct information and complete data. This made the UN adapts "a data revolution" because many developing countries (especially those which have witnessed security crises) lack very much the complete, correct data where big data may appear as a complementary alternative for official statics particularly for the fact of modern technology reaching all countries with no exception due to the spread of cell phones.

However, it has to emphasize using census and demographic surveys as the basic data source, considering that it does not prevent the necessity to develop the statistical systems through big data; it cannot be an alternative for traditional research and theories.

This could enhance the database to be comprehensive and integrated, contributing to the quality of demographic estimations and predicting the different demographic phenomena for the right decision-making in demographic planning. It is known that big data are not based on population, so it is necessary to look for ways to adapt them with what serves the demographic objective, especially that they widen the questions about demographic matters and develop the approaches used.

Using big data in official statics proposes many questions: Do big data have the ability to provide the official statistical data in a precise time? Do they have the ability to face the high costs of producing official statistical work and to provide data and indicators with the required accuracy and efficiency? According to the quality norms of statistical data in the national statics offices and the recommendations of official statistical principles [3].

The demographic statics also require personal data of individuals, their social and demographic status, their economic activities, their educational and healthy environment, and their tendencies and behaviors towards demographic matters, which are difficult to be under the privacy frame. For instance, the previous US president Obama with the previous British Prime Minister Cameron asked tech companies (Facebook and Google) to collaborate with intelligence to track terrorists on social media, which provoked human rights organizations for what is called privacy violation [4].

In spite that using big data in official statics remains far from organizations and statistical bodies' aspirations, but there are leading experiments do not stop on that, and there are many illustrating examples like Estonia, which used mobile-installed tracking devices to improve the touristic statics, and Canada which used GPS devices and Australia which used satellite images for official statistical purposes, whereas in Netherland there are some attempts to use social media means for official statistical purposes too.

Since the UN declaration of 'Data Revolution,' it has been monitored many regional and international seminars and conferences to include big data as a new important source of official statics, the most important one was the Paris and Bangkok meetings (April 23–25, 2013) to unite the efforts of statistical organizations in using big data for data production and to make a united system to classify their different types.

At the 59th World Congress of Statics (China, 25–30 August 2013), big data received a wide interest where participants presented some of these data sources like cell phones, energy consumption, electronic payments, and the ways of processing them through cloud computing.

The last was Arabian, the 4th International Conference of Middle East College in big data and Smart Cities (Oman, Jan, 15. 2019).

The aim was to demonstrate the importance of big data and smart cities in improving the living conditions and planning for cities.

According to the UN economic committee's report, the team is interested in cell phone data for official statics through the partnership between statistical offices and cell phone companies. So it has been the subject where two scientists conducted a work based on analyzing mobile data from 11 countries (EU, Middle East, and Indonesia) using a specific application, preparing, and modeling data by practical examples like applications of typical demographic mobility, migration patterns, transportation, and moving because of diseases and tourism and deleting the population complexes data immediately.

One of the best examples of using big data in official statics is the experience of Belgium. In a comparative study between Belgium census and mobile data extracted from the largest operator there (Proximus) for estimating the demographic density in terms of quality and accuracy that represent an approximate number of inhabitants located in a specific area, where night periods represent a good indicator of residency despite some sort of shortages and mistakes (like having more than one mobile for the same person). It has been found that the correlation coefficient between the density recorded by the census and the one recorded by mobile data equals 0.85 [5]. To access, in the future, better results, it has to develop ways of exploiting mobile data that treat problems they face.

## 21.4 THE IMPORTANCE OF BIG DATA IN OFFICIAL STATICS

Big data may allow enriching official statics and treating the problems resulted from the traditional ways of collecting them, in the following:

1. **Velocity**: field surveys require from preparation to making to showing results a long time, so from a timing perspective, there are annual bulletins, others quarterly or monthly. For instance, concerning the annual ones, we find them late for years what influences all that relies on these data.
2. **Accuracy and Quality:** This shortage of delaying data announcements may influence their quality and importance. In addition, because surveys are built on the samples that do not give exact results only if they have been selected carefully, this affects in return the estimation and planning process of inhabitants' needs.

Generally, we may come out with the importance of big data in official statics in the following points:

- Reducing the costs of data collection;
- Reducing the burden on participants;
- Diversity, richness, accumulation, and immediate availability of data;
- Accuracy and quality.

## 21.5 THE CLASSIFICATION OF BIG DATA SOURCES THAT MAY BE USED IN OFFICIAL STATICS

Big data sources used in producing official statics vary, where the statistical committee's report in 2014 set them in the following resources:

- The emerging sources from one program management: electronic medical records, insurance records, and bank records.
- Sources of tracking devices: satellite images, road sensors, climate sensors.
- Sources of tracking devices like cell phones and GPS.
- Sources of behavioral data like the number of making research on the internet and page viewing.
- Sources of data associated with opinions like comments on social media.

## 21.5.1   PROBLEMS OF USING BIG DATA IN OFFICIAL STATICS

It is not easy to deal with big data as a source of official statics as it is for traditional sources because it faces many problems. The most critical ones are:

- There is a huge amount of stored big data that exposes how to treat and exploit them.
- So the problem of analysis for the absence of sufficient ways and technical programs to do that.
- These data may not be of the needed quality; for instance, they do not cover a specific type of variables or do not include the target society units that give approximately the same problems of traditional studies (bias, missing values, and data quality and completeness).
- They require techniques and ways, first from searching to collecting to analyzing that develop accordingly with their size and diversity.
- Dealing with this type of data confronts property and privacy rights what means the necessity of a legal frame.

In general, one may summarize the problems of using big data in official statics as it is mentioned in the UN statistical committee of 2014 in the legitimate aspects, privacy aspects, financial, management, and technological aspects, then finally the methodological aspects.

## 21.5.2   REQUIREMENTS OF USING BIG DATA IN OFFICIAL STATICS

Big data help to establish an effective information system for the official statics through:

- Exchanging interactive information on the internet for statistical information.
- Continuous work on developing programs that allow recording, processing, collecting, and analyzing these data.
- Providing specialists with big data with the needed skills and technological techniques.
- Benefiting from developed countries through training specialists on official statics derived from big data.
- Monitoring, auditing, frequent, and continuous data collection (immediate to work).

## 21.6   USING BIG DATA FOR SUSTAINABLE DEVELOPMENT PURPOSES

In statistical terms, to achieve sustainable development goals, it must provide data according to age, socio-economic characteristics, and residency location to make sure that all social groups, especially the weak ones, have achieved their desired goals.

These statics have to be available locally, not just nationally, and they have to cover all domains and to be continuously updated [6]. This is very found in big data. However, still the weak groups' problem is there, which prevents achieving these goals simply because these groups do not possess modern technology like the internet and mobiles.

In this context, the statistical committee created the scientific team concerned with using big data for official statics purposes in its 45th session in 2014. The working team is responsible for providing a strategic vision about big data for official statics purposes and directing the program to meet the purposes of indicators designed in the sustainable development plan [7].

In Bogota announcement, it came in "A World Depends on Statics: Harnessing Data Revolution for Sustainable Development Purposes" which recommended using legally these data to produce statics, particularly those measuring the sustainable development indicators:

- Using technology, innovation, and analysis to establish a system that includes networks of data innovation to benefit from data, data research.
- Capacity building, sources correlated with capacity building, technology transfer, data literacy, and source mobilization through innovative financing mechanisms in partnership with private sectors.

For this, work teams were created in the different fields of data usage: satellite, cell phones, social media means, survey devices, and other teams for developing related skills.

It was also discussed in the UN international Forum on Data in Dubai (Oct, 22–24, 2018), especially the point of big data and using them in developing real-time data of migration through recorded calls. It also dealt with the attention of raising agricultural productivity using satellite data to identify the fertile lands from the non-ones.

There are many attempts that relied on big data in facing problems; these later are difficult to process with the traditional ways in different sectors for sustainable development, including:

- The study of the UN food security program used mobile data to assess food security. The results showed that it is possible to use phone balances as an alternative indicator of the market's food spending level [8].
- Very frequent monitoring data were used in Brazil for calculations related to water.
- Satellite images were used in Colombia for mobility and agriculture statics.
- Thanks to satellite imaging at night, Sudan was able to complete data related to poverty.

Using big data has allowed identifying the geo-locations extracted from police calls in tracking traffic accidents in Ouagadougou, without a statistical survey, and estimating the deaths and disorders resulting from these accidents [9].

The Global Pulse Initiative is a remarkable initiative that paid attention to big data. It is an innovative initiative of the UN Secretary-General on data sciences. The initiative enhances the awareness of big data opportunities in the sustainable development and the humanitarian work context; it also aims to develop analytical, highly effective solutions and make them available for the United Nations and the Governmental partners through its network of innovative centers of data sciences.

They are Pulse Labs in Jakarta, Indonesia, and Kampala, Uganda, and the United Nations Headquarters in New York. They aim to reduce restrictions that prevent widening the scope and relying on it [10].

## 21.7 CONCLUSION

Big data are a part of the data revolution, which may improve official statics for achieving sustainable development goals; to fight hunger and poverty, achieving equality, and supporting partnerships.

This is all through the exploitation of different available sources of "data" in terms of quantity, timing, accuracy, and diversity. Therefore, statistical offices should develop their systems in terms of means, techniques, and skills despite what conveys as challenges due to the fast produced data, in return the slow availability of scientific methods in their use, configuration, analysis, and legitimate privacy.

## KEYWORDS

- **big data**
- **data**
- **official statistics**
- **sustainable development**

## REFERENCES

1. The Commission of Sciences and Technology for Development, (2008). *The Commission of Sciences and Technology for Development.* www.un.org (accessed on 24 November 2020).
2. www.un.org. (s.d.). (accessed on 22 October 2020).
3. Abu Dhabi Statistics Center, (2013). Available at: www.scad.ae (accessed on 22 October 2020).
4. Ahmed, A. B., (2017). *Big Data, its Characteristics, Opportunities, and Strength.* http://www.alfaisal-scientific.com:?p=2093 (accessed on 22 October 2020).
5. https://economie.fgov.be (accessed on 22 October 2020).
6. Cazabat, (2018). The United Nations facing big data: How to use new data sources to optimize the development programs of international organizations. *COSSI Journal.*
7. Nations, U., (2018). *The Economic and Social Council: Using Big Data for the Purposes of Official Statics.* The statistical committee.
8. Big Data for Development in Action: The Global Pulse Project Series, UNPG, (2015). *Global Pulse Initiative.* https://www.unglobalpulse.org/2015/07/big-data-for-development-in-action-the-global-pulse-project-series/ (accessed on 24 November 2020).
9. Bedecarrats, F., & J. C., (2016). *Revolution of the Data and Stakes of Statics in Africa,* 288.
10. www.onu.org. (s.d.) (accessed on 22 October 2020).

# CHAPTER 22

# The Initiatives of the UN to Improve the Quality of Big Data and Support the Sustainable Development Goals for 2030

ZAHIA KOUACHE[1] and NADIA MESSAOUDI[2]

[1]Senior Lecturer, Faculty of Economics, Business, and Management Sciences, University of Djilali Bounaama, Khemis Miliana, Algeria, E-mail: z.kouache@univ-dbkm.dz

[2]Assistant Professor, Faculty of Economics, Business, and Management Sciences, University of Djilali Bounaama, Khemis Miliana, Algeria, E-mail: n.messaoudi@univ-dbkm.dz

## ABSTRACT

Sustainable development has become the focus of discussion for many countries. In order to enforce the 2030 sustainable development plan and create a prosperous future for citizens everywhere, the United Nations has pursued reliable, comprehensive, and recorded data by introducing new projects and identifying structures to improve financing and funding and to obtain better data for sustainable development. These data are known as big data. The study concluded that big data management increases job opportunities, improve education and healthcare, thus achieving a better future in line with the sustainable development goals 2030.

## 22.1 INTRODUCTION

Big data has a unique characteristic that distinguishes it from data derived from traditional sources, and it is characterized by its large quantity, ultra-high-speed, and its diversity, in a way that requires effective activities that require cost-effective and creative work to understand it and to use it better in the decision-making process.

By virtue of its large size, it is difficult to deal with and manage it in ways and traditional applications as they are grouped into databases and conduct searches and participation, analysis, comparison, and conclusions, as they are conducted hundreds and hundreds of processes that of course need a large number of devices and people to do them.

The use of big data does not depend on enterprises and businesses but extends to many areas, including energy, education, health, and large scientific projects. In 2015 the world began to work on the scheduling of development work, known as the Sustainable Development Goals, and requires the realization of these objectives. Complementary work focuses on social, economic, and environmental challenges and on participatory and inclusive development that is left behind.

But there is still a lack of critical data that is necessary to develop national, regional, and global development policies. Many Governments lack to access adequate data relating to their people and apply that is for the poorest and most marginalized individuals. The groups that state leaders should focus on to achieve eliminating extreme poverty and eliminating dangerous emissions to the environment, and change them by 2030 through an inclusive process.

The following are the main questions:

*How can big data science contribute to supporting the 2030 Sustainable Development Goals? What are THE UN initiatives to improve the quality of big data?*

Research objectives: Through this research paper, we aim to:

- Identify the nature and structure of big data, their characteristics, and sources.
- Identify the value of big data and its ability to support different institutions and support sustainable development goals.
- Develop a road map to overcome the challenges of investing in big data in decision-making processes and for various development purposes.

## 22.2  BIG DATA, ITS CHARACTERISTICS, SOURCES, AND THE CHALLENGES OF ITS USE

### 22.2.1  BIG DATA DEFINITION AND FEATURES

The term "big data" refers to a very large, organized, or unorganized data set that is so complex that it requires more sophisticated processing systems than traditional data processing systems.

The characteristics of big data are as follows [1]:

1. **Size:** Is the volume of data extracted from a source, which determines the value and volume of data to be classified as big data, and by 2020 the electronic space will contain approximately 40,000 megabytes of data ready for analysis and extraction (Figure 22.1).

## The Digital Universe: 50-fold Growth from the Beginning of 2010 to the End of 2020

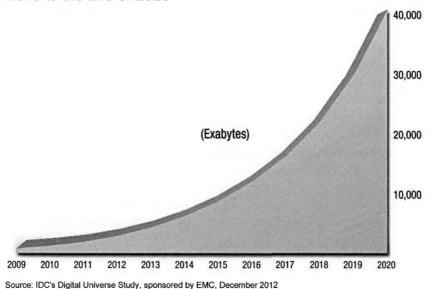

Source: IDC's Digital Universe Study, sponsored by EMC, December 2012

**FIGURE 22.1**    Data volume increases from 2010 to 2020.
*Source:* Ref. [1].

2. **Diversity:** The diversity of extracted data that helps users, whether researchers or analysts, choose the right data for their research field and includes data structured in unstructured databases and data such as photos, clips, audio recordings, videos, SMS messages, and records calls; and map data (GPS) requires time and effort to be configured in a suitable format for processing and analysis.

3. **Speed:** The speed of production and extraction of data to cover the demand where speed is a critical element in decision-making based on this data, which is the time it takes from the moment this data reaches the moment the decision is made based on it.

4. **Veracity:** It means the accuracy, validity, and novelty of such data as one in three executives do not trust the data offered to them.

### 22.2.2   BIG DATA SOURCES

Big data sources vary as [2]:

- Sources arising from the management program: a government or non-governmental programs, such as electronic records of beneficiaries, publishers, employees, client libraries, and visits by beneficiaries. For example, insurance records, bank records, and medical records of patients.
- Commercial or transaction-related sources: data arising from transactions between two entities, for example, credit card transactions and transactions conducted via the Internet, including mobile devices.
- Sensor network sources: such as satellite imagery, road sensors, climate sensors, and air pollution, for example: tracking data from mobile phones and GPS sourcing.
- Behavioral data sources: for example, the number of searches on the Internet for a product, service, or another type of information, and the number of times a page is viewed on the Internet.
- Sources of opinion-related data: for example, comments, and opinions on social media such as Facebook and Twitter.

### 22.2.3   THE CHALLENGES OF USING BIG DATA

The use of large data in many challenges is mainly the following categories:

- **Legislative Aspects:** It refers to data access.
- **Privacy Aspects:** Managing public trust and accepting the reuse of statements and then linking them to other sources.
- **Financial Means:** i.e., the call that can be made to take statements in the interests of their benefits.
- **Management-Related Aspects:** For example, policies and guidance on data management and protection.
- **Methodological Aspects:** The quality of the data and the appropriateness of statistical means.
- **Technological Aspects:** i.e., issues relating to information technologies.

## 22.3   THE SECOND AXIS: SUSTAINABLE DEVELOPMENT AND ITS PLAN FOR 2030

Economic development in its current form has imposed huge burdens on the economy, in particular, the deterioration of environment and treatment, so sustainable development has been a burden for it in order to reduce the severity of these costs and take into account the environmental dimension when building development plans.

### 22.3.1   SUSTAINABLE DEVELOPMENT: DEFINITION AND DIMENSIONS

The World Environment Committee knew it in 1987 as follows: 'It is the development that meets the needs of the present with the inability of future generations to satisfy their needs, or the process of change in which the exploitation of resources is integrated into the direction of investments and the direction of technological development and change. Institutional and enhance sought-after and future energies to satisfy human needs and aspirations" [3].

Sustainable development has three dimensions related to the economic dimension, the social dimension, as well as the environmental dimension, as follows [4]:

1. **Economic Dimension:** Sustainability on the economic side means continuity and maximizing economic well-being for as long as possible by providing the best quality of human well-being such as food, housing, health, and education.
2. **The Environmental Dimension:** The environmental dimension of sustainable development focuses on taking into account environmental boundaries, so that each ecosystem has certain limits that cannot be exceeded by consumption and depletion, but if these limits are exceeded, this leads to environmental degradation and on this basis, the limits must be set for consumption and growth—population, pollution, water depletion, etc.
3. **The Social Dimension:** The social dimension of sustainable development focuses on the essence of development and its ultimate goal through caring for social justice, combating poverty, and providing social services to all those who need it, in addition to ensuring

democracy through the participation of peoples in Making a decision with transparency.

### 22.3.2  SUSTAINABLE DEVELOPMENT PLAN 2030

In 2015, countries adopted the 2030 Sustainable Development Plan and their 17 Sustainable Development Goals and 169 associated goals, which are integrated and indivisible goals to move towards sustainable development. Each state has full-lasting sovereignty and freely exercises all its wealth.

Its natural resources, economic activities, sustainable development, and objectives are integrated and indivisible. They are universal in nature and comprehensive in their application, taking into account the different living realities of each country and its capacities and level of development, as well as respecting national policies and priorities. The goals are considered to be endeavors of the same global character that aspires to be achieved, with each government setting its own national goals guided by the level of global ambition, taking into account national circumstances. Each government must also decide on ways to integrate these ambitious global goals into national planning processes, policies, and strategies, and it is important to take into account.

The link between sustainable development and other ongoing processes related to it in the economic, social, and environmental fields [5].

However, basic data on many targets are still not available, which requires increased support to strengthen data collection and capacity building in the Member States in order to lay the foundations for data nationally and globally.

On the one hand, on the other hand, the gap in data collection needs to be filled in order to help measure progress better, particularly for targets that do not have clear digital features, and this is what we will address in the next axis, outlining the UN's efforts to collect big data to support the Sustainable Development Goals of 2030.

### 22.4  GLOBAL PULSE INITIATIVE AND THE UN GLOBAL DATA FORUM

The 17 Sustainable Development Goals represent the world's most ambitious plan to promote the sustainable development of the planet and its population, and big data will give new impetus to mobilize support, build partnerships to improve the quality of data and statistics, and enhance countries' ability to harness effectively. Data for the implementation of the Sustainable Development Plan-2030.

### 22.4.1   GLOBAL PULSE INITIATIVE

Global Pulse is an innovative initiative of the UN Secretary-General on Data Science, an initiative that inspires awareness of the opportunities that big data offers for sustainable development and humanitarian work and aims to develop high-level analytical solutions.

The impact it has made available to the United Nations and government partners through its network of innovative data science centers, Bolus Labs in Jakarta, Indonesia, Kampala, Uganda, and UN headquarters in New York, aims to reduce barriers to accreditation and expand the scope.

To make safe and responsible use of data, Global Blues has established a data privacy program, one part of which relates to ongoing research into uses that protect the privacy of big data for human and development purposes, and global blues have been established:

1. **Data Privacy Advisory Group:** This group consists of privacy experts from legislators, the private sector, and academia who have engaged in a dialogue on critical issues related to big data, and the group has provided advice on the development of privacy tools and guidelines on the scope of the United Nations system.
2. **Benefits, Damages, and Risk Assessment Tool:** This tool works in two phases to understand the risks associated with big data and includes guidelines to help practitioners assess the proportionality of risks and damages in data-driven projects [6].

### 22.4.2   FIRST UN DATA FORUM

In January 2017 in Cape Town, South Africa, the Global Action Plan for Sustainable Development Data was launched at the opening of the first UN Data Forum, and calls for governments, political leaders, and the international community to commit to key actions in six strategic areas:

- Coordination and leadership;
- Innovation;
- Modernization of national statistical systems;

- Publishing data on sustainable development;
- Building partnerships; and
- Mobilizing resources.

The Cape Town Global Action Plan on Sustainable Development Data, to be adopted by states at the UN Statistical Committee meeting with input from the global statistical community and relevant experts, was prepared, and during the Forum, 1,400 experts from more than 100 countries held panel discussions data systems laboratories and interactive presentations with participants from governments, national statistical offices, the private sector, academia, international organizations, and civil society organizations.

The First United Nations Data Forum concluded its work in Cape Town, South Africa, with the launch of a global data plan to improve people's living conditions, including new ideas and solutions to enhance cooperation, resources, and policies, and the UNITED Arab Emirates has been announced to host the Forum of Nations United 2nd Data in Dubai in 2018.

### 22.4.3   SECOND UNITED NATIONS GLOBAL FORUM

Under the umbrella of the United Nations in Jumeirah's city in Dubai, the Forum brings together leading data producers and users to contribute to innovative initiatives that contribute to improving the quality of data on health, migration, refugees, education, income, environment, and human rights.

The Forum included more than 80 sessions, including interactive visual data presentations and traditional panel discussions that allow participants to interact and exchange ideas and constructively.

The United Nations estimates that 90% of the world's data was created in the expansion of new data sources resulting mainly from mobile, digital, and space technologies that offer broad opportunities for innovative solutions that should be integrated with formal data mechanisms and structures [8].

Plenary sessions and panel discussions focused on key issues within six themes agreed by the Organizing Committee [9]:

- Adoption of new approaches to national capacity-building and their financing to improve data;
- Combining data sources and continuing to integrate non-traditional data sources;
- Making use of data and statistics to represent and speak out, without neglecting anyone;

- Increase data and knowledge of statistical knowledge and enhance communication, including data published in the press;
- Build confidence in data and statistics by applying data and governance principles to new and existing data sources and implementing open data practices;
- Review progress in the implementation of the Cape Town Global Plan of Action agreed at the first Global Data Forum and in addressing emerging challenges.

Improving the use of data and statistics will be critical to achieving the transformative vision for a better future for man and the Earth, as set out in the 2030 Agenda for Sustainable Development.

## 22.5   CONCLUSION

The big data field has grown rapidly with the spread of the Internet and social networks revolution and has become a target for many large companies and governments due to its importance both in the current and future period, and scientists believe that this field will lead the world to a huge revolution that may change its future.

Especially if big data is harnessed to achieve a better future for human societies in line with the 2030 Sustainable Development Goals, including food security, health, and education; the other 17 goals simultaneously address climate change and environmental protection.

- ➢ **Research Results:** The sustainable development goals will be required by 2030 to:
  - Renewed focuses on and applied funding resources to improve the quality of life of the inaccessible.
  - Significant improvements in data collection and analysis.
  - Data is a vital process for decision-making and accountability, and while big data analysis is common in the private sector, development experts can adopt similar techniques to gain real-time insights into human well-being and interventionist aid measures to improve the targeting of vulnerable groups.
  - The purpose of big data management is to achieve results that can be achieved, increase employment opportunities, and improve health and nutrition, nutrition, and gender equality without neglecting anyone.

- ➢ **Prospects of Research:**
  - Bridging the digital divide and creating knowledge-based societies.
  - Ensure data privacy.
  - Encourage and foster innovation to fill graphic gaps.

## KEYWORDS

- **big data**
- **global pulse**
- **sustainable development**

## REFERENCES

1. Al-Bar, A. M. *Big Data and its Application* (pp. 2, 3). King Abdulaziz University. Available at: https://www.awforum.org/index.php/en/component/k2/item/190 (accessed on 24 November 2020).
2. Zainab, B. T., & Al-Rubai, S. B. I., (2018). New roles of information specialist to deal with big data. *Journal of Information and Technology Studies* (p. 7). Specialized Libraries Association, Gulf Arab Branch Qscience, Hamad Bin Khalifa University Publishing House.
3. Carl, C. *Running Some Theoretical Issues: Environment, Employment, and Development* (p. 46). League of Arab States Press, Cairo.
4. Murad, N., (2010). Sustainable development and its challenges in Algeria. *Journal of Communication, 26*, 135–136.
5. General Assembly, (2015). Seventieth Session, Agenda Items 15,116. *Transforming Our World: The 2030 Agenda for Sustainable Development* (pp. 7, 13, 17). United Nations.
6. http://www.un.org/en/sections/issues-depth/big-data-sustainable-development/index.html (accessed on 22 October 2020).
7. http://www.un.org/ar/sections/issues-depth/big-data-sustainable-development/index.html (accessed on 22 October 2020).
8. Emirates News Agency, (2018). https://al-ain.com/article/united-nations-data-forum-dubai-kicked-monday (accessed on 22 October 2020).
9. *Launch of the 2018 UN Global Data Forum to Promote Innovative Data Solutions.* Dubai, Press Release. https://undataforum.org/WorldDataForum2018.pdf (accessed on 22 October 2020).

# CHAPTER 23

# Big Data and Its Role in Achieving the Sustainable Development Goals: Experiences of Leading Organizations

KAMEL MAIOUF and ACHOUR MEZRIG

*Department of Sciences Economy, Commercial, and Management Sciences, Hassiba Ben Bouali University of Chlef, Pb. 02000, Algeria, E-mail: m.kamel@univ-chlef.dz (K. Maiouf), m.kamel@univ-chlef.dz (A. Mezrig)*

## ABSTRACT

The technological revolution that the world is witnessing in recent times and in various fields led to the emergence of the term big data, where this concept is one of the concepts that the leading global organizations (Facebook, Google, Amazon) seek to pay attention to it widely, especially with the huge profits it makes. Big data can be adopted to achieve the well-being of societies and benefit them in achieving the Sustainable Development Goals through their use in many areas such as energy, education, and health. This study aims to provide the contribution of big data to achieving the Sustainable Development Goals under the head of the leading organizations in this field.

## 23.1 INTRODUCTION

Globally, the big data technology market and related services are expected to grow at a rate of about seven times faster than the ICT market as a whole. According to Gartner, a leading IT research company: Big data is now at a peak in inflation, and it is expected to reach the productive peak about 5 to 10 years, after a few years of cloud computing.

Big data imposed itself as a new reality as a result of technological developments, which made it possible to know the quality of data that was not available in the past. What was accompanied by this technological explosion is the emergence of modern technological media, social media, and various applications that have made it possible to know huge data in each area. Most of the leading business organizations (Google, Facebook, and Amazon) have paid wide attention to this data due to their profits. It should be mentioned that the use of big data does not concern only the business area, but extends to many other areas including energy, education, and health.

Big data can be adopted to achieve the well-being of communities, their use in providing aid to vulnerable groups, and new sources of data and new technologies. If implemented effectively and responsibly, it can contribute to sustainable development goals. Therefore, this study aims to analyze the following question:

> How can big data contribute to achieving Sustainable Development Goals in order to lead business organizations?

## 23.2   THE GENERAL FRAMEWORK OF BIG DATA

Big data has become one of the most popular terms, especially with the huge technological explosion in all areas of everyday life, so what do we mean by big data? What are its most important characteristics and its source?

### 23.2.1   DEFINITION AND CHARACTERISTICS OF BIG DATA

Big data is defined as data generated through the use of a lot of digital devices, computers, and everything that is connected to the Internet at any given moment. For example, millions of people around the world using mobile phones, send text messages or e-mail, or to view digital content on the network, etc. Through these different usages, a great amount of data can be created.

All these activities leave a digital impact, and then this information together constitutes what is known as big data.

Big data is characterized by its large size, the speed in terms of data-generating and updating, and the multiplicity of body or shape. This means that big data needs a more advanced technology in storage and processing to convert this data from its raw image to value knowledge, which is the essence of benefit.

In 2011, the Mackenzie Global Institute defined big data as any collection of data that is larger than the ability of traditional database tools to capture, store, manage, and analyze that data [3].

The most widely used definition of big data in the industrial field, the definition of Gartner in 2012 one of the leading companies in the field of IT research and consulting, which says that: big data is a high-volume, fast, and diverse information asset, and requires innovative and cost-effective forms of processing Information enables to enhance vision and decision-making [4].

From the above, big data can be defined as a collection of very large, complex data that is difficult to process and managed using methods. Traditional accepted applications are grouped into databases and analyzed, stored, and used in the time of need.

The methods of processing big data, in addition to the huge volume of data produced, stored, and made available under the umbrella, are characterized by other characteristics different from traditional data, or which are stored neatly and coordinated as databases. Experts believe that the most important characteristics of data are [1]:

1.  **Volume:** Experts estimate that by 2020 the Internet will contain approximately 40,000 zettabytes of data ready for analysis and debriefing.
2.  **Velocity:** To process a small set of data stored in databases, companies were analyzing each data set separately and sequentially until they were all completed. But as the volume of data swelled, the need for special systems to ensure the speed of big data analysis was urgently needed. That need led to the creation of special techniques for processing such data.
3.  **Variety:** The number of people who have been displaced by the war has increased.

### 23.2.2  BIG DATA FORMS, SOURCES, AND IMPORTANCE

Organizations and institutions are currently acquiring additional data from their operational environment at an increasingly rapid pace, such as [4]:

- **Web Data:** Web behavior data can be captured at the customer level, such as page views, searches, comments, and purchase slot, where it can enhance performance in areas such as best-in-class presentation, form planting, customer segmentation, and targeted ads.
- **Text Data:** e-mail, news, Facebook posts, and is one of the largest types of application of most data on a large scale.
- **Time and Location Data:** It makes GPS and mobile phone technologies as well as Wi-Fi information from time and location information a growing source of data. And at the individual level, many organizations come to achieve the power of knowing when and where customers are, and it is equally important to look at time and location data is on a collective level, and many individuals open their time and location data more openly, and many interesting applications begin to appear. So big data from time and location should be treated with extreme caution for their privacy.
- **Smart Grid and Sensor Data:** Sensor data is collected at present from cars, oil pipes, mill turbines, collected at an extremely high frequency, and as sensor data provide strong information on the performance of engines and machines, it can diagnose problems more easily and faster to develop procedures.
- **Social Network Data:** Social networking sites or social networking, the most important of which is Facebook, LinkedIn, Twitter, and Instagram. It is possible to do link analysis to reveal a network of a particular user. The analysis of the social network can give ideas about ads that may attract selected users. This is done by looking not only at the interests mentioned by customers personally but also knowing what is important within the circle of their friends or colleagues.

According to a report on the coordination of the Economic Commission for Europe, entitled "What the big data means for official statistics" on 10 March 2013, a summary of the data sources was developed, as these data are now generated automatically and continuously in digital format in many different ways, and in general, big data sources can be classified as follows [4, 5]:

1. **Sources Arising from the Management of a Program:** Whether it is a government or non-governmental programs, such as electronic medical records, hospital visits, insurance records, bank records, and food banks.

2. **Commercial or Transaction-Related Sources:** Data arising from transactions between two entities, for example, credit card transactions and transactions conducted via the Internet, including by mobile devices.

3. **Source of Sensor Networks:** Such as satellite imagery, road sensors, climate sensors, and air pollution.

4. **Tracking Device Sources:** Such as tracking data from mobile phones and GPS.

5. **Behavioral Data Sources:** Such as the number of internet searches for a product, service, or any other type of information, and the views of a page on the Internet.

6. **Sources of Opinion-Related Data:** Such as comments and opinions on social media such as Facebook and Twitter.

### 23.2.3   BIG DATA CHALLENGES

The use of big data in statistics poses many challenges, which is mainly in the following categories:

- **Legislative Aspects:** The committees are also required to provide the information.
- **Aspects of Privacy:** Managing public confidence and accepting the reuse of data.
- **Financial Aspects:** The costs involved in taking data for the benefits achieved.
- **Management Aspects:** For example, policies and guidance on data management and protection.
- **Methodological Aspects:** The quality of the data and the appropriateness of statistical means.
- **Technological Aspects:** i.e., issues relating to information technologies.

### 23.3   THE THEORETICAL FRAMEWORK FOR SUSTAINABLE DEVELOPMENT

Sustainable development is one of the most important goals that organizations and governments want to reach through the use of different methods and strategies.

Since the official emergence of sustainable development, many researchers and international organizations have been exposed to its definition because of the importance of this topic, thus gaining great global attention, especially after the emergence of the Brundtland report entitled "Our Common Future" Prepared by the World Environment and Development Committee in 1987.

The first definition of sustainable development was established in this report as "development that meets current needs without compromising the ability of future generations to meet their needs [6]," and the report issued by the World Resources Institute included the limitation of 20—a broadly worded definition of sustainable development.

The report divided these definitions into four groups: economic, environmental, social, and technological. Economically means sustainable development of developed countries to reduce energy consumption, and resources for underdeveloped countries means employment. Resources to raise the standard of living and reduce poverty. Socially and humanly means seeking to stabilize population growth and raise the level of health services, especially in rural areas, but at the environmental level.

It means protecting natural resources and optimizing the use of agricultural land and water resources, and finally, it means moving society to the age of clean industries that use environmentally clean technology and produce minimal polluting, heat produces minimal pollution, heat-trapping, and ozone-harmful gases [7].

FAO defines sustainable development as managing and protecting the natural resource base and guiding technical and institutional change in a way that ensures that the human needs of present and future generations are achieved and continue to be met so that this development seeks to protect the earth.

Water and plant and animal genetic sources are not harmful to the environment and are technically appropriate, economically appropriate, and socially acceptable. Social, environmental, and technological and contribute to maximizing growth in the previous four systems [8]. The dimensions of sustainable development are:

1. **The Economic Dimension:** The economic dimension of sustainable development represents the current and future effects of the economy on the environment and raises the issue of selecting, financing, and improving industrial techniques in the field of the employment of natural resources [7].

2.  **The Social Dimension:** The social dimension of sustainable development focuses on the fact that human beings are at the center and core, both as a means and a goal, and therefore concerned with: Social justice, equality, and the fight against poverty by supporting national action plans and programs [6]; and by providing and improving the level of major social services to all those in need. Sustainable development is particularly characteristic with this dimension, as the human dimension of the narrow sense makes growth a way for social integration, and the process of development in the political selection and this choice must be accepted by all Choose away between the people selections as they are between the countries.

3.  **The Environmental Dimension:** In their approach to sustainable development, environmentalists focus on the concept of 'environmental boundaries,' which means that each natural ecosystem has certain limits that cannot be exceeded by consumption and depletion, and any encroachment on this natural capacity means that the ecosystem is degraded. So sustainability from an environmental perspective always means setting limits to consumption, population growth, pollution, poor production patterns, water depletion, deforestation, and soil erosion [9].

Sustainable development, through its mechanism and content, seeks to achieve a set of goals that can be summarized as follows [10]:

-   To achieve a better quality of life for the population, where sustainable development is pursued through the planning process and implementation of development policies to improve the quality of life of members of society economically, socially, and psychologically by focusing on the qualitative aspects of growth in a fair and secure manner.
-   Respect for the natural environment, where sustainable development focuses on the activities of the population, deals with natural systems and their content on the basis of human life, or simply a development that understands the sensitive relationship between nature and the built environment, and develops this relationship into a relationship of integration and harmony.
-   To enhance the awareness of the population of existing environmental problems, where their sense of direction belongs, and encourage them to actively participate in finding appropriate solutions through their

participation in the preparation, implementation, follow-up, and evaluation of sustainable development programs and projects.

- Achieving the rational exploitation and use of resources, where such development treats natural resources as limited resources, thus preventing depletion or destruction and rationally using and employing them.
- Linking modern technology to society's objectives, where it tries to employ modern technology to serve the goals of society by educating the population about the importance of various technologies in the field of development and how to use them to improve the quality of life.

Therefore, human beings are an essential element of sustainable development, as they seek to meet their needs and organize their lives so that they can deal with natural resources with knowledge and wisdom.

## 23.4 BIG DATA AND ITS ROLE IN ACHIEVING SUSTAINABLE DEVELOPMENT DUE TO THE EXPERIENCES OF LEADING BUSINESS ORGANIZATIONS

Due to the technological explosion and the rapid development in various areas of life, the emergence of the vast amount of data being produced, stored, and easily accessible from multiple locations, this big data, if exploited effectively, will contribute to sustainable development.

### 23.4.1 LEADING ORGANIZATIONS IN BIG DATA ANALYSIS

According to statistics, by 2020, it is expected that smartphone generates more than 2GB of data per month, and this rapid growth in data production is due to the proliferation of Internet-connected devices and systems. With the shift from the analogy to digital technologies, and the rapid and rapid use of digital media by organizations and individuals, we also find that the share of the production of irregular data through social media, videos, and images is currently greater than the share of regular data [11].

Due to the increase in activities, the world around us currently produces more than 1.7 trillion bytes of data per minute, of which some centers according to Intel, the volume of data produced by the world from the beginning of the Internet age until 2003 is estimated at more than 5 Exabyte's

(Exabyte equivalent to 1 billion gigabytes), it refers to more than 500 times during 2012 to 2.7 Zettabyte is equivalent to 1,000 billion gigabytes, and experts predicted that this trend would increase by the end of 2018.

Because of this huge volume of data, the term big data began spreading, it has increased in speed so that it is difficult to process them now using a single program or independent device or using traditional data processing applications, and here technology companies have begun to develop new auxiliary software and hardware through which they can help analyze that big data, although you wonder these examples [1]:

1.  Amazon analyses and processes millions of operations daily to meet the wishes of its customers, in addition to responding to the queries of more than half a million sellers per day, so Amazon has the three largest databases in the world;

2.  Wal-Mart, on the other hand, processes more than 1 million businesses per hour, which are stored in databases containing more than 2.5 petabytes, which is 167 times the content of books in the United States Library of Congress;

3.  WhatsApp has more than 450 million users, trading more than 10 billion messages and 400 million images daily.

In addition, there are other leading companies in the field of big data analysis, like [3]:

- The Great Hydroshock has 150 million sensors that provide data 40 million times per second, and there are approximately 600 million collisions per second, but we only deal with less than 0.001% of the sensor current data, the flow of data from all four collider experiments represents 25 petabytes.

- Facebook handles 50 billion images from its user base, and the credit card protection system FICO Falcon Credit Card Fraud Detection System protects 2.1 billion active accounts worldwide.

- Windermere real estate uses anonymous GPS signals from nearly 100 million drivers to help new home buyers determine their driving times to and from work during different times of the day.

- IBM says that we produce 2.5 quintillion bytes of data every day (Quintillion is the number one followed by 18 zeros). This data comes from everywhere, such as climate information and comments posted on social media digital images, videos, and sales and purchase transactions.

## 23.4.2   THE CONTRIBUTION OF BIG DATA TO ACHIEVING THE SUSTAINABLE DEVELOPMENT GOALS

In 2015, with the approval of 193 countries, the United Nations set global sustainable development goals, and the Sustainable Development Goals will continue from 2015 to 2030 with 17 goals: no to poverty, no hunger, good health, quality education, gender equality, clean and healthy water, renewable energy and affordable reasonable, good jobs, innovative, and good infrastructure, reducing inequality, cities, and sustainable communities, responsible use of resources, climate action, climate action, sustainable land use, peace and justice, and partnership for sustainable development.

The UN Secretary-General's independent advisory panel on harnessing the data revolution for sustainable development made specific recommendations on how to achieve these goals:

- Promoting and encouraging innovation to fill the gaps.
- Mobilizing resources to overcome inequality between developing and developed countries, and to overcome inequality between peoples rich in data resources and people slackening them.
- Creating leadership and coordination to allow the data revolution to play its full part in achieving sustainable development.

At the first United Nations Global Data Forum, held in January 2017, more than 1,400 public and private sector data users and producers, as well as policymakers, members of academia and civil society, came together to explore ways to harness the power of data for sustainable development.

The contribution of big data to achieving the Sustainable Development Goals shows [12]:

- **Poverty Eradication:** The patterns of the spending provided by cellular can identify income indicators.
- **Ending Hunger Altogether:** Tracking food resource prices online can help to monitor food security in real-time.
- **Healthy Water and Sanitation:** Sensors connected to water pumps can track access to clean water.
- **Clean and Affordable Energy:** Smart metering allows utility companies to increase the flow of electricity, gas, water, and ensure adequate supply during peak periods.

- **Underwater Life:** Marine ship tracking data can reveal illegal or unregulated fishing activities.
- **Sustainable Cities and Communities:** Remote satellite sensing can track encroachment on land and public spaces such as parks and forests.
- **Peace, Justice, and Strong Institutions:** Analysis of emotions on social networks can reveal public opinion about effective governance, the provision of public services, or human rights.
- **Partnerships to Achieve Goals:** Partnerships that combine statistics, mobile data, and internet data can lead to an immediate understanding of the interconnectedness of today's world as a small village.
- **Production and Responsible Consumption:** Internet search patterns or e-commerce transactions can reveal the pace of the transition to energy-saving products.
- **Life on Land:** Social media monitoring can support disaster management with immediate information on victims' locations and the effects of firepower and fog.
- **Economic Growth:** Global postal traffic patterns can provide indicators of economic growth, remittances, trade, and GDP.
- **Industry, Innovation, and Infrastructure:** Data from GPS devices can be used.
- **Climate action:** By combining satellite imagery, eyewitness testimonies, and open data, it can help track deforestation.
- **Good Health and Well-Being:** Mapping mobile phone users' movement can help predict the spread of infectious diseases.
- The committees are responsible for the work of the Committee, and the committees are responsible for the work of the Committee.
- **Gender Equality:** Analysis of financial transactions can reveal spending patterns for men and women.

Big data can highlight previously unseen disparities in society. For example, women and girls who often work in informal sectors or at home suffer from social constraints on their movement, as well as marginalization in the decision-making process.

As an example of big data and its role in achieving sustainable development, we find a global plus, an innovative initiative of the Un Secretary-General on data science, which aims to raise awareness of the opportunities offered by big data in relation to sustainable development purposes.

Humanitarian work aims to develop high-impact analytical solutions. It provides solutions to the United Nations and government partners through its network of innovative data science centers. The Bols Labs in Jakarta, Indonesia, Kampala, Uganda, and the UNITED Nations headquarters in New York, and to ensure access to big data views across multiple industries has not been able to work with the private sector to activate the concept of data philanthropy, which is intended to make safe and responsible use of partnership data in humanitarian action and development purposes.

For example, Global Blues in 2016 partnered with a Twitter social media platform, and people around the world daily exchange tweets that promise hundreds of millions of dollars in dozens of languages, which contain real-time information on issues such as the cost of food. Jobs, access to medical care, quality of education, and reports on natural resources will be made available to United Nations development and humanitarian agencies.

The possibility of converting public data into actionable information to help countries around the world, one example of such partnerships is the big data for the social good initiative, which uses the big data available from cellular networks to address humanitarian crises, including natural disasters and epidemics. There is also the climate action data initiative, which connects researchers around the world with tools and data available from leading companies to provide data-based solutions to climate issues [12].

There is a growing recognition that the success of the SDGs will depend on the ability of governments, companies, and civil society organizations to harness data in the decision-making process, and the key is to invest in building innovative data systems that rely on new sources of real-time data.

For sustainable development, big data through smartphone applications helps school workers and community education register students and teachers' attendance on a transparent and instantaneous basis and follow up more easily with dropout students.

Especially for reasons that can be overcome through interventions based on Informed by community education workers, this information can be automatically fed to statistics that educators can use to track progress in key areas and allow big data through smartphone applications to record patient information in each visit.

This information can go directly to public health statistics that health officials can use to monitor disease outbreaks or the need to strengthen technical personnel; such systems are able to provide a real-time record of important events, including births and deaths.

Even the use of so-called oral autopsy help to determine the cause of death. As part of electronic medical records, information can be used for future visits or to remind patients of the need to follow up on visits or medical interventions. By increasing the effectiveness of data use massive collected during the provision of services, the global energy system becomes more efficient and less polluting. And vital services such as health and education become more effective and easier to obtain.

This data accelerates sustainable development by improving decision-making. This is only a first step as the same methods should also be used to gather some key indicators that measure progress in achieving the Sustainable Development Goals. The United Nations Solutions Network will help sustainable development in support of the New Global Partnership through the creation of a new specialized data network on sustainable development. This will bring together many data scientists, thinkers, and academics from many sectors to form a center of excellence in the field of data. This is to say that it will be possible to transform data into real progress in sustainable development [13].

## 23.5   CONCLUSION

Through our analysis of the topic of big data and its role in achieving the Sustainable Development Goals under the experiences of leading organizations, we have found that the emergence of big data was due to the technological explosion in various fields and the accompanying emergence of the Internet, computers, and technological applications.

This data has become a high-level interest by leading business organizations (Facebook, Google, Amazon.) because it works to improve the standard of living and improve the well-being of society on the one hand; and to drive innovation and advancement in all areas and achieve sustainable development on the other hand. And despite all this, data still faces many challenges, most notably the privacy of users, etc.

## KEYWORDS

- **big data**
- **leading organizations**
- **sustainable development**

## REFERENCES

1. Arabic Channel. *Era-Big Data-How-Benefited-World-Ones*? https://www.alarabiya. net/ar/qafilah/2018/02/28/%D8%B9%D8%B5%D8%B1-%D8%A7%D9%84% D8%A8%D9%8A%D8%A7%D9%86%D8%A7%D8%AA-%D8%A7%D9% 84%D8%B6%D8%AE%D9%85%D8%A9-%D9%83%D9%8A%D9%81-%D- 8%A7%D8%B3%D8%AA%D9%81%D8%A7%D8%AF-%D8%A7%D9%84%D8 %B9%D8%A7%D9%84%D9%85-%D9%85%D9%86%D9%87%D8%A7%D8%9F (accessed on 22 October 2020).
2. Islam, D., & Alaa, Z., (2018). *Big Data Analysis-Big Data Practical Experiences* (p. 1). Egyptian Center for Economic Studies, Egypt.
3. Mohamed, H., (2013). *Big Data Profile*. https://www.tech-wd.com/wd/2013/07/24/ what-is-big-data/ (accessed on 22 October 2020).
4. Zineb, B. T., & Al-Rayai, S. B. I., (2018). New roles of information specialist to deal with big data. *Journal of Information and Technology Studies* (p. 5). Specialized Libraries Association, Arab Gulf Branch.
5. Economic and Social Council, (2013). *Big Data and Statistical Systems Modernization* (pp. 8, 9). United Nations.
6. Hossin, Y., & Iman, A., (2016). The role of social responsibility for organizations in the embodiment of sustainable development. *The 13ᵗʰ International Conference on the Role of Social Responsibility for Small and Medium Enterprises in Strengthening the Strategy of Sustainable Development-Reality and Bets* (p. 9). Chlef University.
7. Achour, M., et al., (2019). Social responsibility as a mechanism for achieving sustainable tourism development in Algeria as part of the 2030 tourism development guideline. *The 14ᵗʰ International Forum on Citizenship and Social Responsibility Behaviors for Business Organizations in the Country* (pp. 7, 8). Chlef University, Algeria.
8. *The Concept of Sustainable Development*. https://www.seo-ar.net (accessed on 22 October 2020).
9. *How to Measure Sustainable Development*. https://www.startimes.com/?t=6931062 (accessed on 22 October 2020).
10. Baala, T., & Youcef, A. M., (2016). Social responsibility of the organizations the new path to achieving the sustainable development goals. *The 13ᵗʰ International Conference on the Role of Social Responsibility for Small and Medium Enterprises in Strengthening the Strategy of Sustainable Development-Reality and BETS* (pp. 11, 12). Chlef University.
11. Ministry of Information and Communications Technology, (2014). *Big Data: Balancing Advantages and Risks*. Qatar.

12. *Big Data for Sustainable Development.* http://www.un.org/ar/sections/issues-depth/
big-data-sustainable-developmen (accessed on 22 October 2020).

13. *Data Cycle for Sustainable Development.* https://www.env-news.com/in depth/
articles/21341/ (accessed on 22 October 2020).

# Index

For Product Safety Concerns and Information please contact our EU
representative GPSR@taylorandfrancis.com Taylor & Francis Verlag GmbH,
Kaufingerstraße 24, 80331 München, Germany


Printed and bound by CPI Group (UK) Ltd, Croydon, CR0 4YY
08/05/2025
01864402-0001